Student Solutions Manual for

STATISTICS FOR BUSINESS AND ECONOMICS

James R. Lackritz

San Diego State University

Debra Olson Oltman

Oral Roberts University

Brooks/Cole Publishing Company

Pacific Grove, California

Brooks/Cole Publishing Company
A Division of Wadsworth, Inc.

© 1991 by Wadsworth, Inc., Belmont, California 94002. All rights reserved. No part of this book may be be reproduced, stored in a retrieval system, or transcribed, in any form or by any means — electronic, mechanical, photocopying, recording, or otherwise — without the prior written permission of the publisher, Brooks/Cole Publishing Company, Pacific Grove, California 93950, a division of Wadsworth, Inc.

Printed in the United States of America

10 9 8 7 6 5 4 3 2 1

ISBN: 0-534-14433-0

PREFACE

This manual provides worked out solutions to all odd-numbered exercises in the main text. You should be aware that many of the problems can be solved in more than one way. Therefore, our solutions and explanations may not be unique.

We caution you to use this manual judiciously. It is very easy to study a solution and to assume that the solution is not difficult to achieve. Instead, we urge you to use this manual as a check to your work and to assist in finding and correcting your mistakes, and not as an excuse to say that your homework is done. There is no substitution for a lot of hard work.

James R. Lackritz

Debra Olson Oltman

CONTENTS

Chapter 1		Statistics for the Business World	1
Chapter 2		Organizing and Presenting Data	3
Chapter 3		Descriptive Measures	11
Chapter 4		Probability	21
Chapter 5		Probability Distributions	27
Chapter 6		Sampling Distributions and Sampling Procedures	38
Chapter 7		Estimation	43
Chapter 8		Hypothesis Testing	49
Chapter 9		The F Distribution and Analysis of Variance	57
Chapter 10		Simple Regression and Correlation	73
Chapter 11		Multiple Regression and Correlation	84
Chapter 12		Time Series and Forecasting	88
Chapter 13		Index Numbers	97
Chapter 14		Decision Making	103
Chapter 15		Chi-Square Tests	116
Chapter 16		Nonparametric Tests	127

CHAPTER 1

STATISTICS FOR THE BUSINESS WORLD

1.1 Answers will vary 1.3 Answers will vary 1.5 Answers will vary

1.7 Answers will vary. Some possible answers are asking price, square footage, number of bedrooms, number of bathrooms, number of stories, lot size.

1.9 Answers will vary. Some possible answers are: (a) Temporary residents, people without phones or unlisted numbers (b) Paid employees (c) Students, poor people (d) working people (e) mothers (f) unemployed

1.11 (a) effectiveness of drug(yes or no) (b) discrete (c) Population is *all people*. Sample is the 1000 people chosen. (d) the parameter is the percent of all people for whom the drug is effective. The statistic is the percent of the sampled people for whom the drug is effective. The statistic is 97%. The parameter is unknown.

1.13 discrete quantitative: b,h discrete qualitative: c,f continuous: a,d,e,g,i,j

1.15 No. The population is all Americans 18 and over.

1.17 (a) descriptive (b) inferential 1.19 Answers will vary

1.21 Cylinders are not consistent and it would be impossible to measure all cylinders, since all would end up broken and there would be none to use.

1.23 Answers will vary. Some possible responses are (a) How good was it to begin with (b) How many people were asked to state a preference (c) What does *superior* mean (d) How many doctors were asked for a recommendation (e) How many cases were tried.

1.25 (a) A sample may be unusual and contain persons whose responses differ from those of most people. (b) The sample may purposely or inadvertently exclude a group of people.

1.27 They exclude people who (1) have unlisted numbers, (2) have no phone, and (3) work hours when the polls are being conducted

1.29 (a) all packing boxes of the type used by the company (b) the packing boxes used by the company in the study (c) amount of pressure required for collapse (d) amount of pressure required for collapse for all boxes in the population (e) amount of pressure required for collapse for the boxes in the sample

1.31 descriptive: b,c inferential: a,d,e,f

1.33 (a) All tape cannot be tested (b) This is constantly changing (c) This is currently unknown and consistently changing (d) advertising must be changed continually, and may not be able to reach all potential clients (e) This may be constantly changing.

1.35 By chance, it may include members that do not accurately represent the population in some way.

1.37 Answers will vary

1.39 (1) pieces of information (2) a field of study, and (3) a measure that describes some quantifiable characteristic of a population.

1.41 The result could, by chance, be the same that would have occurred had statistical procedures not been abused.

1.43 Answers will vary

1.45 Intentional abuses are the result of design; unintentional abuses are the result of ignorance.

1.47 No. the possibility of error using valid procedures is always present.

CHAPTER 2

ORGANIZING AND PRESENTING DATA

2.1

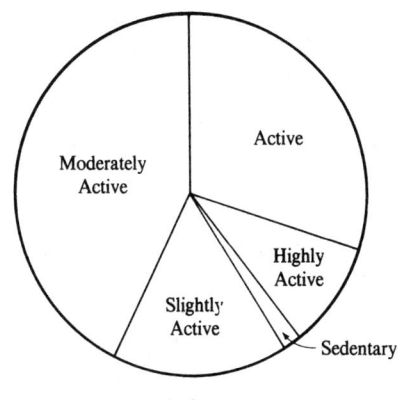

Pie Chart

Sedentary $= \frac{4}{260}(360) = 5.5°$
Slightly Active $= \frac{42}{260}(360) = 58.2°$
Moderately Active $= \frac{111}{260}(360) = 153.7°$
Active $= \frac{78}{260}(360) = 108°$
Highly Active $= \frac{25}{260}(360) = 34.6°$

Bar Graph

Bar Chart

2.3   482 674 773 877 897 920 935 981 1253 1582

2.5 As total gross sales are $1,008,000, each slice should be proportional to the totals. Therefore, Tennis = $(143/1008)360 = 51°$. Ski = $(288/1008)360 = 103°$, Athletic Shoes = $(195/1008)360 = 70°$, Golf = $(22/1008)360 = 8°$, and Clothing = $(360/1008)360 = 128°$. Pie graph sectors should measure $51°(I), 103°(II), 70°(III), 8°(IV), 128°(V)$.

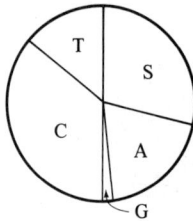

$T = \text{Tennis} = \frac{143}{1008}(360) = 51°$
$S = \text{Ski} = \frac{288}{1008}(360) = 103°$
$A = \text{Athletic Shoes} = \frac{195}{1008}(360) = 70°$
$G = \text{Golf} = \frac{22}{1008}(360) = 7°$
$C = \text{Clothing} = \frac{360}{1008}(360) = 129°$

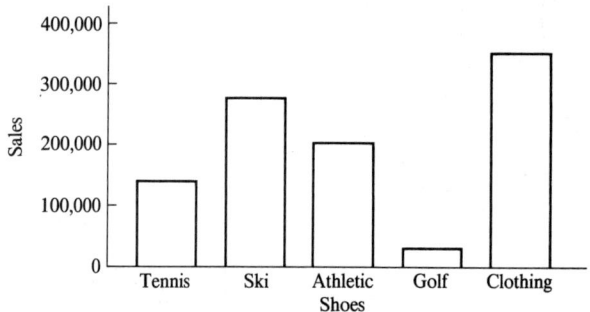

2.7

2.9 (a) A perfect graph will be impossible, as there is no limit on the final graph. If we wanted to assume that the final class has a limit of 600, then the figure below would suffice.

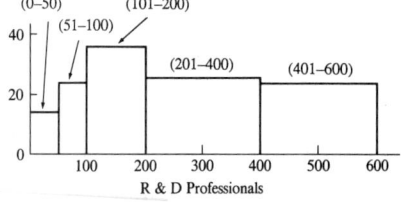

(b) If we use the 2^k rule, with n=123, we would have k=6 or 7. With only 5 classes, we should consider adding an extra class or two, and generating equal class widths.

(c) The weight of the graph is higher on the right side, and it appears to have generated some skewness with this non-symmetric distribution.

2.11 (a) Transportation – 35 Tourism and Entertainment – 50 Oil and Gas – 10
    Manufacturing – 40 Health – 10 Service – 30 Other – 25

(b) $35/200 = .175$, $50/200 = .25$, .05, .2, .05, .15, .125

(c) This is all but 65, or $135/200 = .65$, or 65%.

2.13 (a)

Class Limits	Class Midpoints	f	r
115 – 137	126	1	.022
92 – 114	103	1	.022
69 – 91	80	9	.200
46 – 68	57	4	.089
23 – 45	34	14	.311
0 – 22	11	16	.356
		45	1.00

(b)
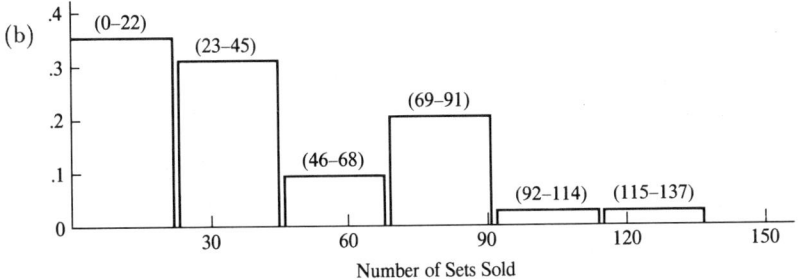

2.15 $2^k \leq 150$. k=7. R = 61/7 = 8.7 → 9.

Class Limits	Class Midpoints	f	r
54 – 62	58	3	.020
45 – 53	49	9	.060
36 – 44	40	18	.120
27 – 35	31	20	.133
18 – 26	22	42	.280
9 – 17	13	33	.220
0 – 8	4	25	.167
		150	1.00

2.17 (a) number of calls. discrete.

Class Limits	Class Midpoints	f	r
35 – 40	37.5	8	.12
29 – 34	31.5	10	.15
23 – 28	25.5	18	.27
17 – 22	19.5	23	.34
11 – 16	13.5	7	.10
5 – 10	7.5	<u>1</u>	<u>.01</u>
		67	.99 (slight deviation from 1.00 due to rounding)

(c)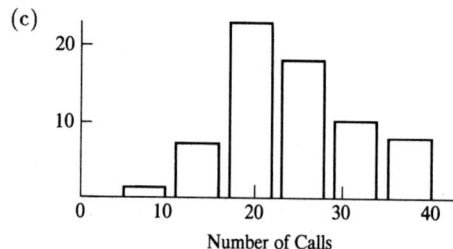

2.19 answers will vary

2.21 (a) 35+50+20+25+5=135

(b) (20+25+5)/145=.345 or 34.5%

(c) Yes. The younger companies have a higher frequencies. Your finance professor might tell you that younger companies tend to have higher growth rates.

2.23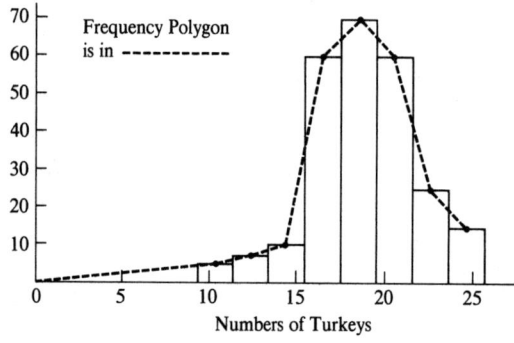

2.25

Class Boundaries	Class Midpoints	f	r
20.005 – 24.005	22.005	2	.051
16.005 – 20.005	18.005	5	.128
12.005 – 16.005	14.005	6	.154
8.005 – 12.005	10.005	9	.231
4.005 – 8.005	6.005	<u>17</u>	.051
		39	

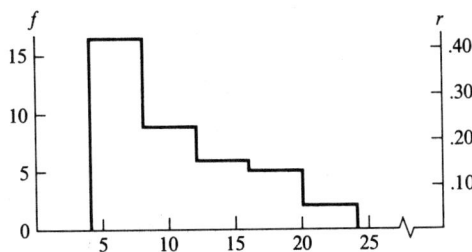

2.27

Class Boundaries	Class Midpoints	f	F
19.05 – 24.05	21.55	2	25
14.05 – 19.05	16.55	7	23
9.05 – 14.05	11.55	11	16
4.05 – 9.05	6.55	<u>5</u>	5
		25	

(b)

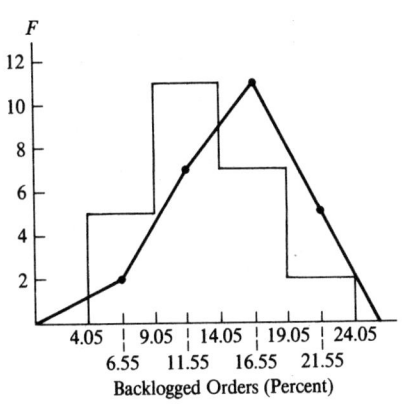
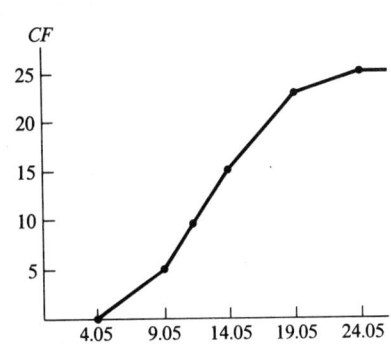

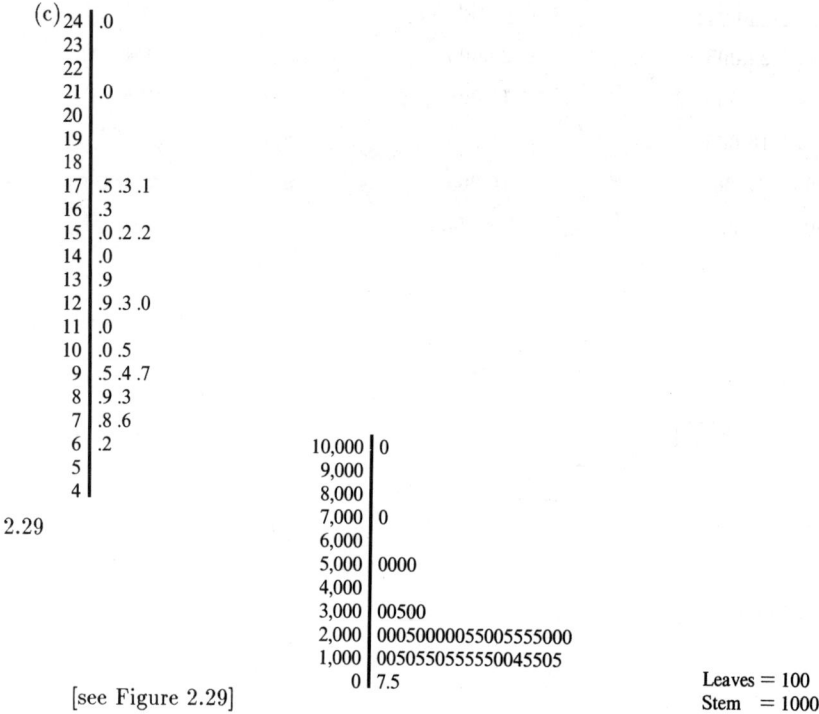

2.29 [see Figure 2.29]

The distribution is heaviest in the 1000 and 2000 stems, and then decreases as the stems get larger.

2.31 $.23N = 46$. $N = 46/.23 = 200$.

2.33 (a) 50%(10 squares out of 20) (b) 10%(2 out of 20) (c) A vertical frequency scale

2.35 See previous figures throughout chapter.

2.37 Increasing the number of intervals leads to more precision but sacrifices simplicity.

2.39 (a) – (b) Results will vary (c) Number that appears on the die – discrete.

2.41 (a) 25+30=55 (b) 30+10+10+10=60 (c) (10+20)/(10+20+25+25+25+30+30+10+ 10+10) = 30/195=.15 or 15%.

2.43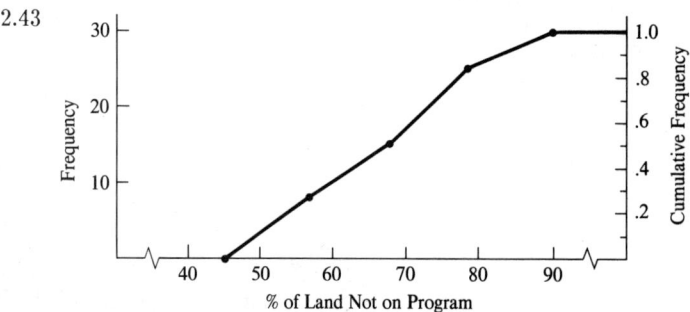

2.45

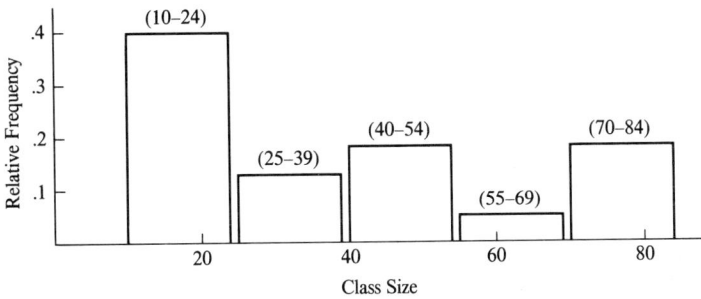

2.47 (a) 10+40+60+30+20=160 (b) (60+300/160=.5625 or 56.25% (c) 150(all but 10)

2.49 (a) 27 (b) 11 (c) 16 (d) 6 (e) 3.2 (f) .10 (g) whole values, tenths, hundredths

2.51

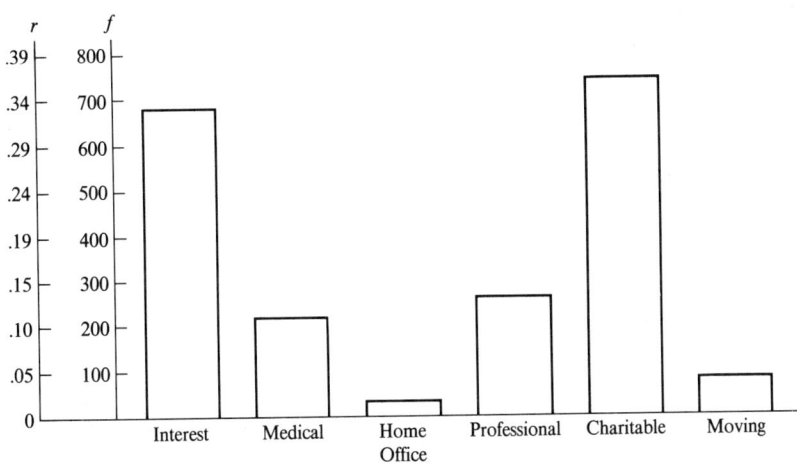

2.53

Class Boundaries	Class Midpoints	f	F
140.35 – 145.35	142.85	5	44
135.35 – 140.35	137.85	7	39
130.35 – 135.35	132.85	8	32
125.35 – 130.35	127.85	13	24
120.35 – 125.35	122.85	<u>11</u>	11
		44	

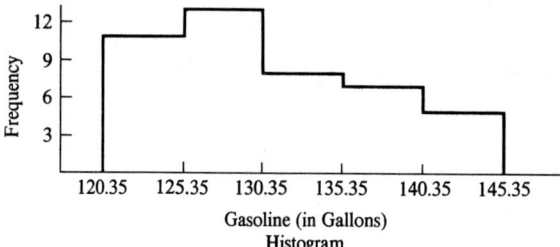

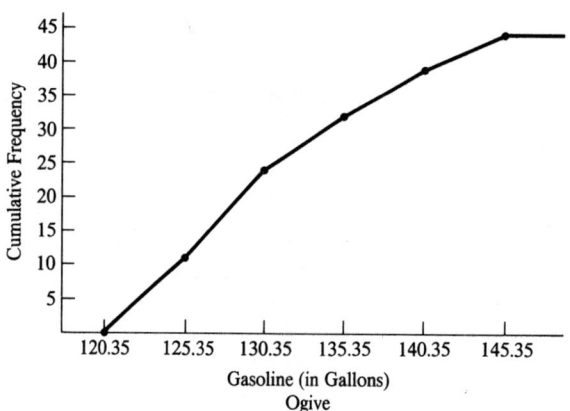

2.55
```
140 | 3.2 0.0 2.1 4.6 5.3 1.0
130 | 7.1 9.4 3.3 7.8 9.4 3.2 2.4 2.4 2.1 0.4 0.0 3.7 9.4 2.0 8.6
120 | 5.4 0.7 2.8 4.5 0.6 7.9 6.3 7.8 8.5 9.6 3.7 9.3 9.6 0.5 2.8 3.7 8.3 9.4 1.3 7.5 7.2 5.0 5.0
```
2.57   answers will vary

CHAPTER 3

DESCRIPTIVE MEASURES

3.1 (a) $\bar{X} = 29.5/6 = 4.92$; median = 4.95; all are mode
(b) $\bar{X} = 28.4/6 = 4.73$; median = 4.2; mode = 4.0
(c) $\bar{X} = 32.4/6 = 5.40$; median = 5.35; mode = 4.1
(d) $\bar{X} = 90.3/18 = 5.02$; median = 4.8; mode = 4.1

3.3 (a) $\mu = 1440/15 = 96$
(b) $\Sigma(X - \mu) = 4 - 13 - 8 - 15 - 13 + 0 + 9 + 12 - 18 + 6 + 1 + 17 + 30 - 2 - 10 = 0$.

3.5 $\mu = 1105/15 = 73.67$; median = 74; mode = 75.

3.7 $\bar{X} = 10/8 = 1.25$ median = 1 mode = 1

3.9 (a) gm = $(3 \times 9 \times 8 \times 4 \times 11)^{1/5} = 6.2457$
(b) gm = $(2 \times 7 \times 4 \times 8)^{1/4} = 4.6007$
(c) gm = $(2.83 \times 9.71 \times 3.11 \times 2.79 \times 15.41)^{1/5} = 5.1646$
(d) gm = $(3.4 \times 7.1 \times 9.8)^{1/3} = 6.1847$

3.11 (a) $\Sigma X = 24$
(b) $\Sigma X^2 = 0^2 + 3^2 + 9^2 + 4^2 + 7^2 + 1^2 = 156$
(c) $(\Sigma X)^2 = 24^2 = 576$
(d) $\Sigma(X - 4) = -4 + (-1) + 5 + 0 + 3 + (-3) = 0$
(e) $\Sigma X - 4 = 24 - 4 = 20$
(f) $\Sigma(X - 4)^2 = (-4)^2 + (-1)^2 + 5^2 + 0^2 + 3^2 + (-3)^2 = 60$
(g) $\Sigma(X - \mu) = \Sigma(X - 4) = 0$
(h) $\Sigma(X - \mu)^2 = \Sigma(X - 4)^2 = 60$
(i) $\Sigma X^2 + 3 = 156 + 3 = 159$
(j) $\Sigma(X - \mu)^2 / N = 60/6 = 10$

3.13 Sample 1: R = 36 − 11 = 25, $\Sigma X = 152$, $\Sigma X^2 = 3396$, $\overline{X} = 152/8 = 19$, mean deviation = $(5+1+0+8+17+8+7+4)/8 = 50/8 = 6.25$

$$s^2 = \frac{3396 - \frac{(152)^2}{8}}{7} = 72.7 \quad s = \sqrt{72.7} = 8.52$$

Sample 2: R = 108 − 23 = 85, $\Sigma X = 473$, $\Sigma X^2 = 37061$, $\overline{X} = 473/7 = 67.57$, mean deviation = $(20.43+21.43+4.57+12.57+20.57+44.57+40.43)/7 = 164.57/7 = 23.51$

$$s^2 = \frac{37061 - \frac{(473)^2}{7}}{6} = 849.95 \quad s = \sqrt{849.95} = 29.15$$

Sample 3: R = 0 − (−25) = 25, $\Sigma X = -65$, $\Sigma X^2 = 1055$, $\overline{X} = -65/6 = -10.83$, mean deviation = $(3.83+2.83+.17+14.17+3.17+10.83)/6 = 35/6 = 5.83$

$$s^2 = \frac{1055 - \frac{(-65)^2}{6}}{5} = 70.17 \quad s = \sqrt{70.17} = 8.38$$

Sample 4: R = 3.7 − (−4.7) = 8.4, $\Sigma X = -1.2$, $\Sigma X^2 = 50.4$, $\overline{X} = -1.2/5 = -0.24$, mean deviation = $(1.86+3.94+.86+3.24+4.46)/5 = 14.36/5 = 2.872$

$$s^2 = \frac{50.4 - \frac{(-1.2)^2}{5}}{4} = 12.53 \quad s = \sqrt{12.53} = 3.54$$

3.15 (a) $\Sigma X = 6{,}455$, $\Sigma X^2 = 5{,}053{,}787$, $\overline{X} = 6455/10 = 645.5$,

$$s^2 = \frac{5053787 - \frac{(6455)^2}{10}}{9} = 98{,}564.94$$

(b) It would still be 98,564.94

3.17 All observations are identical. The average, median, and mode will all be the same as the observations, and the range is 0.

3.19 (a) $\Sigma X = 569$, $\Sigma X^2 = 15{,}049$, $\mu = 569/25 = 22.76$

(b)
$$\sigma = \sqrt{\frac{15049 - \frac{(569)^2}{25}}{25}} = 9.16$$

(c)
$$s = \sqrt{\frac{15049 - \frac{(569)^2}{25}}{24}} = 9.35$$

3.21 $\Sigma X = 8$, $\Sigma X^2 = 18$, n = 5, $\overline{X} = 8/5 = 1.6$, s^2(defining formula) = $[(0-1.6)^2 + (1-1.6)^2 + (2-1.6)^2 + (2-1.6)^2 + (3-1.6)^2]/4 = 5.20/4 = 1.30$

computational formula: $s^2 = \frac{18 - (8)^2/5}{4} = 5.2/4 = 1.3$

3.23 $\Sigma X = 120.95$, $\Sigma X^2 = 975.5985$, $\overline{X} = 120.95/15 = 8.0633$

$$s^2 = \frac{975.5985 - (120.95)^2/15}{14} = .024167 \quad s = .1555$$

$\overline{X} \pm 2s$ goes from $8.0633 - 2(.1555)$ to $8.0633 + 2(.1555)$, or from 7.7523 to 8.3743. All but 7.73 falls in the interval, or $14/15 = .933$

3.25 $(X+Y)/2 = 4 \quad (X-4)^2 + (Y-4)^2 = 18 \to X=1, Y=7$

3.27 $\Sigma X = n \quad \Sigma X^2 = n^2 \quad s^2 = \dfrac{n^2 - \frac{n^2}{n}}{n-1} = \dfrac{n^2 - n}{n-1} = \dfrac{n(n-1)}{n-1} = n \quad s = \sqrt{n}$

3.29 We will work with the numerators, since both denominators are $n - 1$. $\Sigma(X - \overline{X})^2 = \Sigma X^2 - 2\overline{X}\Sigma X + \Sigma \overline{X}^2 = \Sigma X^2 - 2\overline{X}(n\overline{X}) + n\overline{X}^2 = \Sigma X^2 - n\overline{X}^2 = \Sigma X^2 - n(\Sigma X/n)^2 = \Sigma X^2 - (\Sigma X)^2/n$.

3.31 (a) $16(.25) = 4 \to 4.5$, $Q_1 = 3{,}950$.
  (b) $16(.35) = 5.6 \to 6$, $P_{35} = 4{,}300$.
  (c) $16(.5) = 8 \to 8.5$, $D_5 = 5{,}250$.
  (d) $16(.63) = 10.08 \to 11$, $P_{63} = 6{,}500$.
  (e) $16(.9) = 14.4 \to 15$. $P_{90} = 8{,}100$.

3.33 (a) 19.525 is 3.625% above the mean, or $3.625/2.5 = 1.45$ standard deviations above the mean. Similarly, 12.275 is 3.625 below the mean, or 1.45 standard deviations. With $J = 1.45$, we have $1 - 1/(1.45)^2 = .5244$ or at least 52.44% of the observations in this interval.
  (b) 19.72 is 3.82% above the mean, or $3.82/2.5 = 1.53$ standard deviations above the mean. With $J = 1.53$, we have at least $1 - 1/(1.53)^2 = .5717$ or 57.17% of the observations in this interval.
  (c) 21.4 is 5.5% above the mean, or $5.5/2.5 = 2.2$ standard deviations above the mean. With $J = 2.2$, we have at least $1 - 1/(2.2)^2 = .7934$ or 79.34% of the observations in this interval.

3.35 (a) position 30.5, $P_{60} = .315$ (b) position 38, $Q_3 = .35$
  (c) position 35.5, $D_7 = .335$ (d) position 9, $P_{17} = .21$
  (e) position 13, $P_{25} = .23$

3.37 (a) $R/4 = (9-1)/4 = 2$ (from before, $s = 2.5$)
  (b) $R/4 = (1244 - 453)/4 = 197.75$ (from before, $s = 236.9$)
  (c) $R/4 = (81{,}700 - 35{,}000)/4 = 11675$ (from before, $s = 14271.8$)

3.39 2 mpg, assuming "most" implies within two standard deviations.

3.41 (a) For age, the z-scores for the minimum and maximum values are
$$z = \frac{18 - 34.5}{11.28} = -1.46 \qquad z = \frac{64 - 34.5}{11.28} = 2.62$$
Therefore, all observations are within 2.62 standard deviations of the mean.

For schooling, the z-scores for the minimum and maximum values are
$$z = \frac{0 - 11.66}{2.73} = -4.27 \qquad z = \frac{17 - 11.66}{2.73} = 1.96$$
Therefore, all observations are within 4.27 standard deviations of the mean.

For work experience, the z-scores for the minimum and maximum values are
$$z = \frac{1 - 14.81}{10.59} = -1.30 \qquad z = \frac{45 - 14.81}{10.59} = 2.85$$
Therefore, all observations are within 2.85 standard deviations of the mean.

(b) Yes. These z-scores cover 100% of the observations, more than the minimum numbers guaranteed by Chebyshev's Theorem.

(c) For age, $s \approx (64 - 18)/4 = 11.25$, very close to the actual value.

For schooling, $s \approx 17/4 = 4.25$, off by more than 50%.

For work experience, $s \approx (45 - 1)/4 = 11$, very close to the actual value.

3.43 (a) $[8 + .5(3)]/20 = .475 \quad PR_{23} = 48$

(b) $[13 + .5(2)]/20 = .7 \quad PR_{25} = 70$

(c) $[0 + .5(1)]/20 = .025 \quad PR_{14} = 3$

(d) $[18 + .5(1)]/20 = .925 \quad PR_{34} = 93$

(e) $[5 + .5(1)]/20 = .275 \quad PR_{20} = 28$

3.45 A has a rank of 5.5. Therefore, $PR_A = 4/8 + 1/2(2/8) = 5/8$ or 63%. H also has this rank. B is ranked second, so its $PR = 1/8 + 1/2(1/8) = 3/16$, or 19%. $PR_C = 2/8 + 1/2(1/8) = 5/16 = 31\%$, $PR_D = 6/8 + 1/2(1/8) = 13/16 = 81\%$, $PR_E = 0 + 1/2(1/8) = 1/16 = 6\%$, $PR_F = 7/8 + 1/2(1/8) = 15/16 = 94\%$, and $PR_G = 3.5/8 = 44\%$.

3.47 (a) $z = \frac{5-4}{3.5} = 0.29$ (b) $z = \frac{-7-4}{3.5} = -3.14$ (c) $z = \frac{2-4}{3.5} = -0.57$

(d) $z = \frac{-4.5 - 4}{3.5} = -2.43$ (e) $z = \frac{3.2 - 4}{3.5} = -0.23$ (f) $z = \frac{4.8 - 4}{3.5} = 0.23$

(g) $4 - .15(3.5) = 3.475$ (h) $4 - 1.45(3.5) = -1.075$ (i) $4 - 2.70(3.5) = -5.45$

(j) $4 + .8(3.5) = 6.8$ (k) $4 + 1.5(3.5) = 9.25$ (l) $4 + 2.33(3.5) = 12.16$

3.49 (a) $\Sigma X = 7104$, $\Sigma X^2 = 4252526$, $\mu = 7104/13 = 546.46$

$$\sigma^2 = \frac{4252526 - \frac{(7104)^2}{13}}{13} = 28497.17 \qquad \sigma = 168.81$$

(b) $z = \frac{373 - 546.46}{168.81} = -1.03 \qquad z = \frac{415 - 546.46}{168.81} = -0.78$

$z = \frac{500 - 546.46}{168.81} = -0.28 \qquad z = \frac{400 - 546.46}{168.81} = -0.87$

3.51 (a) For systolic pressure, we have ΣfX = 124.5(13) + 134.5(28) + ... + 204.5(1) = 23726. ΣfX² = 124.5²(13) + 134.5²(28) + ... + 204.5²(1) = 3514041.5.
$\overline{X}$ = 23726/162 = 146.72

$$s^2 = \frac{3514041.5 - (23726)^2/162}{161} = 165.22, \text{ and } s = \sqrt{165.22} = 12.85$$

For diastolic pressure, since the final category in unbounded, we cannot estimate the mean and variance. If we wish to assume that the last category goes from 109.5 − 119.5, we have ΣfX = 64.5(4) + 74.5(8) + ... + 114.5(2) = 14779. ΣfX² = 64.5²(4) + 74.5²(8) + ... + 114.5²(2) = 1359830.5.
$\overline{X}$ = 14779/162 = 91.23

$$s^2 = \frac{1359830.5 - (14779)^2/162}{161} = 71.84, \text{ and } s = \sqrt{71.84} = 8.48$$

(b) For the histogram, again it is not possible for distolic pressure, unless we assume the last category goes from 109.5 − 119.5.

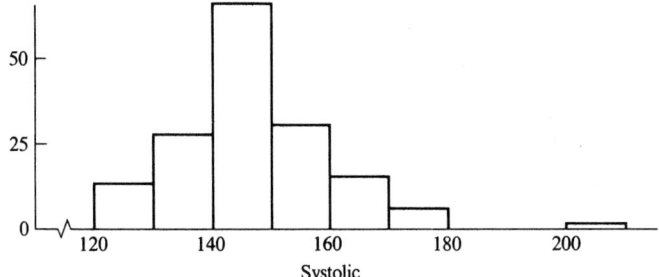

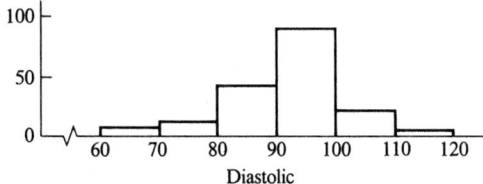

(c) With 162 observations, the median falls between the 81st and 82nd observations, therefore falling in the 139.5 − 149.5 category for systolic pressure, and 89.5 − 99.5 category for diastolic pressure.

3.53 (a) ΣfX = 100556, ΣfX² = 101579734, $\overline{X}$ = 100556/100 = 1005.56

(b) $s^2 = \dfrac{101579734 - (100556)^2/100}{99} = 4693.36$

3.55 $\Sigma fX = 6(30) + 13(71) + ... + 48(41) + 55(9) = 14146$. $\Sigma fX^2 = 36(30) + 169(71) + ... + 2304(41) + 3025(9) = 469794$.

$\overline{X} = 14146/500 = 28.23$

$$s^2 = \frac{469794 - \frac{(14146)^2}{500}}{499} = 142.83$$

3.57 (a) They seem rather high.

(b) n=29 29(.25) = 7.25 → 8, 29(.75) = 21.75 → 22, LH = 2.0, UH = 4.2, Median = 3.3, IQR = 4.2 − 2.0 = 2.2, 1.5IQR = 3.3, UIF = 4.2+3.3 = 7.5, LIF = 2.0 − 3.3 = −1.3(0 for practical purposes), UOF = 10.8, LOF = −4.6.

(c) We have one(9.0) past the upper inner fence and one(13.5) past the upper outer fence. We should check 13.5 to be sure that it was accurate.

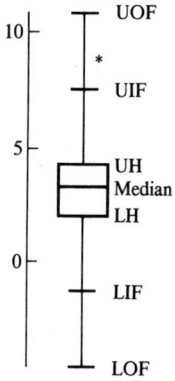

3.59 (a) n=28 28(.25) = 7 → 7.5, 28(.75) = 21 → 21.5 LH = 75, UH = 250, Median = 140, IQR = 250 − 75 = 175, 1.5IQR = 262.50, UIF = 250+262.5=512.5, LIF = 75 − 262.5= −187.5, UOF = 775, LOF = −450.

(b) 3 (c) 3 (750, 800, and 870)

3.61 Yes. This fraction is 128/200 or 64%. This is not surprising. Chebyshev's Theorem makes no guarantee about the number of observations within 1 standard deviation from the mean. While we know that at least 3/4 of the measurements fall within two standard deviations, this leaves plenty of room in the area between the first and second standard deviations.

3.63 (a) $\Sigma fX = 15(350) + ... + 30(80) = 53{,}650$. $\overline{X} = 53{,}650/365 = 146.99$

(b) $\Sigma fX^2 = 15(350^2) + ... + 30(80^2) = 9{,}149{,}500$.

$$s^2 = \frac{9{,}149{,}500 - (53{,}650)^2/365}{364} = 3471.66$$

(c) For the yearly cost, we project the mean to be 365(146.99) = 53,651.35, and the variance to be $365^2(3471.66) = 462{,}511{,}903.50$

3.65 (a) $\Sigma X = 4924$, $\Sigma X^2 = 2{,}287{,}282$, $\overline{X} = 4924/12 = 410.33$, median = 362.5,

$$s^2 = \frac{2287282 - (4924)^2/12}{11} = 24251.61, \ s = 155.74$$

− 16 −

(b) R/4 = (635 − 227)/4 = 408/4 = 102; the difference is 155.74 − 102 = 53.74.

3.67 $\Sigma X = 5.79$, $\Sigma X^2 = 4.3023$, $\mu = 5.79/8 = .724$,
$$\sigma^2 = \frac{4.3023 - (5.79)^2/8}{8} = .014, \sigma = .118$$

3.69 (a) $\Sigma X = 339$ $\Sigma X^2 = 23239$ $\mu = 339/5 = 67.8$ median = 70 all are mode

(b) $\sigma^2 = \dfrac{23239 - \dfrac{(339)^2}{5}}{5} = 50.96$ $\sigma = 7.14$

(c) $z = \dfrac{60 - 67.8}{7.14} = -1.09$ PR = (1 + .5)/5 = .3 or 30%

3.71 (a)

Data Set	X	Y	Z
Mean	5.17	7	69.83
Median	4	7.5	82.5
Mode	4	8	80 and 85

(b) For stocking purposes, we would like to know what happens the most often, so the mode would be appropriate.

(c) Either the mean or median would be appropriate for long-term storage. (d) It is not affected by the extreme value 2.

3.73 (a) 2 ft (or 24 inches) (b) $\mu + \sigma = 24 + 4.5 = 28.5$ in. (c) $\mu + 1.85\sigma = 24 + 1.85(4.5) = 32.33$ in. (d) $\mu - 2.34\sigma = 24 - 2.34(4.5) = 13.47$ in. (e) $\mu - 5\sigma = 24 - 22.5 = 1.5$ in.

3.75 (a) business total is 280, health is 7(45) = 315. Health has more jobs.

(b) business average is 280/5 = 56, more than health's average of 45.

3.77 The total of the 9 terms is 9(25.5) = 229.5. The sum of the first eight terms is 197.7, so the last term is 31.8.
$$\frac{197.7 + X}{9} = 25.5 \Rightarrow X = 9(25.5) - 197.7 = 31.8$$

3.79 (a) $\Sigma X = 710$ $\Sigma X^2 = 21998$ $\overline{X} = 710/25 = 28.4$
$$s^2 = \frac{21998 - \dfrac{(710)^2}{25}}{24} = 76.42 \quad s = 8.74$$

$\overline{X} \pm 1s$ goes from 28.4 − 8.74 to 28.4 + 8.74 or from 19.66 to 37.14. This includes 16 measurements.

$\overline{X} \pm 2s$ goes from 28.4 − 2(8.74) to 28.4 + 2(8.74) or from 10.92 to 45.88. This includes all 25 measurements.

$\overline{X} \pm 3s$ goes from 28.4 − 3(8.74) to 28.4 + 3(8.74) or from 2.18 to 54.62. This includes all 25 measurements.

(b) Yes. Chebyshev's Theorem does not make any guarantees for 1 standard deviation. It promises at least 3/4 of the measurements within two standard deviations of the mean and at least 8/9 of the measurements within two standard deviations of the mean. Our results from (a) are more than these minimum guarantees.

(c) If we started to see too many observations of 46 or more repairs or 10 or fewer repairs, we might be suspicious that something is wrong. We expect to see occasional observations outside of two standard deviations from the mean, but not consistently many.

3.81 n = 15, 15(.25) = 3.75 → 4, 15(.75) = 11.25 → 12, Q_1 = 42100, Q_3 = 71400, median = 62100, IQR = 71400 − 42100=29300, 1.5IQR = 43950, UIF = 71400+43950=115350, LIF = 42100 − 43950 = − 1850(0 for practical purposes), UOF = 159300, LOF = − 45800. There are no outliers. 400,000 would be well outside the UOF and need to be checked for correctness.

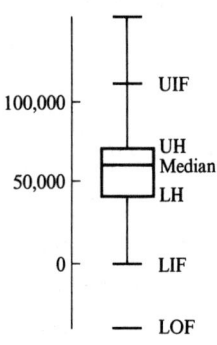

3.83 (a) $\frac{21.7 - 19}{2.75} = 0.98$ (b) $\frac{15.3 - 19}{2.75} = -1.35$ (c) $\frac{12 - 19}{2.75} = 2.55$

(d) When we substitute these values in for k in the term $1/k^2$, we can only use these for $k \geq 1$. Therefore, for 21.7, the maximum fraction is one. For 15.3, we have $1/(-1.35)^2 = .552$, and for 12, we have $1/(2.55)^2 = .154$.

3.85 (a) P_{80} is in position 24.5, at 37 (b) position 15.5, P_{50} = 29

(c) position 8, Q_1 = 22 (d) position 12.5, D_4 = 27.5

3.87 n = 25, 25(.25) = 6.25 → 7, 25(.75) = 18.75 → 19, Q_1 = 15, Q_3 = 27, median = 21, IQR = 27 − 15 = 12, 1.5IQR = 18, UIF = 27+18 = 45, LIF = 15 − 18 = − 3(0 for practical purposes), UOF = 63, LOF = − 21. 47 is the only outlier.

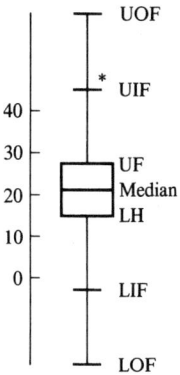

3.89 (a) $\Sigma X = 20{,}350$, $\Sigma X^2 = 51{,}795{,}204$, $\overline{X} = 20350/12 = 2543.75$

$$s^2 = \frac{51795204 - \frac{(20350)^2}{8}}{7} = 4270.21 \quad s = 65.35$$

(b) We expect most of the observations to fall within two standard deviations of the mean, 2543.75 ± 2(65.35), or from 2413.05 to 2674.45.

3.91 .45 = (X+1.5)/50 => X = 21

3.93 $PR_{3.8}$ = (1+.5)/25 = .06 or 6% $PR_{14.4}$ = (21+.5)/25 = .86 or 86% $PR_{5.5}$ = (5+.5[2])/25 = .24 or 24%

3.95 (a) $\Sigma fX = 961$ $\Sigma fX^2 = 17937$ $\mu = 961/58 = 16.57$

(b) $\sigma^2 = \dfrac{17937 - \dfrac{(961)^2}{58}}{58} = 34.73$ $\sigma = 5.89$

3.97 (a) Yes. If $0 < \sigma^2 < 1$, then $\sigma^2 < \sigma$. (b) no

(c) Yes, if all of the terms are the same.

(d) Yes. If $\sigma^2 = 0$ or 1, then $\sigma = 0$ or 1.

3.99 (a) $\Sigma f = 249{,}877{,}000$ (remember frequencies are in thousands)

(b) This cannot be done unless we can place a boundary on the final class. If we assume this to be 100 − 104, and call this to be a population, we have

$\Sigma fX = 2(18{,}409{,}000) + 7(18{,}333{,}000) + ... + 102(57{,}000) = 8{,}715{,}299{,}000$

$\Sigma fX^2 = 2^2(18{,}409{,}000) + 7^2(18{,}333{,}000) + ... + 102^2(57{,}000) = 427{,}419{,}863{,}000$

$\mu = 8{,}715{,}299{,}000/249{,}877{,}000 = 34.878$

$$\sigma^2 = \frac{427{,}419{,}863{,}000 - (8{,}715{,}299{,}000)^2/249{,}877{,}000}{249{,}877{,}000} = 494.02,$$

and $\sigma = \sqrt{494.02} = 22.23$

(c) Assuming the last category at 100 − 104, the histogram is given below.

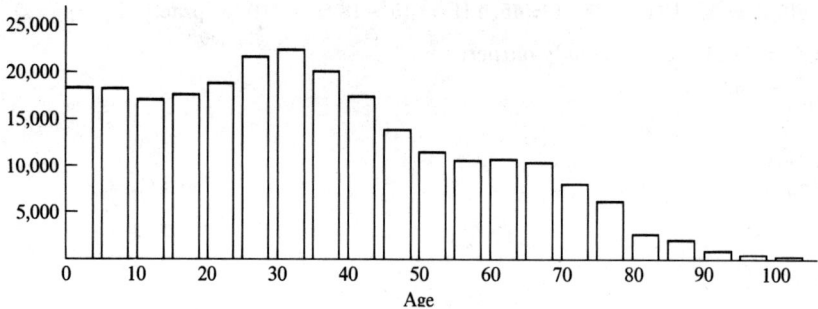

The distribution is skewed with the mean considerably higher than the median, with numerous observations on the high end pulling up the mean.

PROBABILITY

4.1 Empirical   4.3 Empirical: a,c,e Classical: d,f,g Subjective: b

4.5 (a) S = {CCC, CCI, CIC, CII, ICC, ICI, IIC, III} This is the same as tossing three coins.

(b) i. CCC ii. CCI, CIC, ICC iii. CCC, CCI, CIC, ICC iv. CCI, CIC, CII, ICC, ICI, IIC, III

(c) If P(C) = P(I) = .5

4.7 (a) S = {CCC, CCP, CPC, CPP, PCC, PCP, PPC, PPP} This is the same as tossing three coins.

(b) 7/8 (c) 3/8

4.9 (a) $\binom{5}{2} = 10$, using the combinations rule.

(b) Let the men be A, B, and C, and the women be D and E. Then S = {AB, AC, AD, AE, BC, BD, BE, CD, CE, DE}

(c) 6/10 = .6

4.11 $1/\binom{49}{6} = 1/13{,}983{,}816$

4.13 $P(A \cup B) = .33 + .17 - 0$(the groups are mutually exclusive since they are never on special at the same time) = .5

4.15 A = {AB, AC, AD, AE, BC, BD, BE, CD, CE}, B = {AD, AE, BD, BE, CD, CE, DE}.
$P(A \cup B) = 10/10 = 1$. $P(A \cap B) = P\{AD, AE, BD, BE, CD, CE\} = 6/10 = .6$

4.17 (a)

[Venn diagram: A=.20, B=.25, C=.26, D=.06, outside=.23]

(b) $P(B \cup C \cup D) = .25 + .26 + .06 = .57$

(c) $P(\text{Merge}) = P(A \cup B \cup C \cup D) = .20 + .25 + .26 + .06 = .77$

(d) $P(\text{Not Merge}) = 1 - P(\text{Merge}) = 1 - .77 = .23$

4.19 P(A ∪ B) = P(A) + P(B) − P(A ∩ B), so we have .64 = .23 + .59 − X, so X = P(A ∩ B) = .18

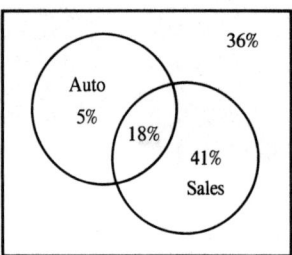

4.21 (a) A and B are mutually exclusive, since if we get the same number on the die, it would be impossible to have them sum to seven. A and C are not mutually exclusive since we can have (6,6). B and C are not mutually exclusive, since we could have (6,1) or (1,6).

(b) A ∩ C means we have the same number on each die AND at least one of the dice has a six, implying two sixes. B ∩ C means the numbers add up to 7, AND at least one die has a 6, leading to a (6,1) or (1,6).

(c) A = {(1,1),(2,2),(3,3),(4,4),(5,5),(6,6)}, B = {(1,6),(2,5),(3,4),(4,3),(5,2),(6,1)}, C = {(1,6),(2,6),(3,6),(4,6),(5,6),(6,6),(6,1),(6,2),(6,3),(6,4),(6,5)}, P(A ∪ B) = 6/36+6/36 = 12/36 = 1/3; P(A ∪ C) = 6/36+11/36 − 1/36 = 16/36 = 4/9; P(B ∪ C) = 6/36+11/36 − 2/36 = 15/36 = 5/12.

(d) P(A ∩ B) = 0; P(A ∩ C) = 1/36; P(B ∩ C) = 2/36.

(e) P(A') = 1 − 6/36 = 30/36 = 5/6; P(B') = 1 − 6/36 = 30/36 = 5/6. P(C') = 1 − 11/36 = 25/36.

4.23 A simple event contains only one member of the sample space. A compound event contains two or more events of the sample space.

4.25 By multiplying the group percentages by the total numbers, we can recreate the table to look as such:

Age	Medicare	Blue Cross	Uninsured	Totals
0 − 9	2100	5063	5028	12,191
10 − 17	685	1768	1417	3,870
18 − 24	1320	2577	6083	9,980
25 − 64	3187	20553	15251	38,991
	7,292	29,961	27,779	65,032

(a) P(18 or older) = (9,980+38,991)/65,032 = 0.75

(b) P(18 or older|uninsured) = $\frac{\text{P(18 or older} \cap \text{uninsured)}}{\text{P(uninsured)}}$ = $\frac{(6083+15251)/65032}{27779/65032}$ = 0.77

(c) P(Blue Cross) = 29,961/65,032 = 0.46

(d) P(BC|25 − 64) = 20533/38991 = 0.53

(e) P(Medicare or Blue Cross) = (7292+29961)/65032 = 0.57

(f) P(BC or 10 − 24) = (29,961+12,191 +3,870 − 5,063 − 1,768)/65,032 = 0.60

(g) P(uninsured ∩ 18 − 24) = 6083/65,032 = .094

4.27 (a) When you add up all of the numbers, there are 77(million) people in the sample. Therefore, we have (.3+1.7+2.2+3.3)/77 = 7.5/77 = .097

(b) (7.5+2.2+5.7+3.3)/(10.6+3.9+7.5+2.2+5.7+3.3) = 18.7/33.2 = 0.563

4.29 (a) 27/80 = .34 (b) 53/80 = .66 (c) 13/80 = .16 (d) 65/80 = .81

(e) 38/80 = .475 (f) 67/80 = .84 (g) 29/80 = .36 (h) 1

4.31 P(D ∩ S') = P(D)P(S'|D) = .63(.13) = .0819

4.33 P(A|B) = $\frac{6/10}{7/10}$ = 6/7; P(B|A) = $\frac{6/10}{9/10}$ = 2/3. No, they are not independent. P(A|B) = 6/7 ≠ P(A) = 9/10.

4.35 Yes. If P(A|B) = P(A), then 1 − P(A|B) = 1 − P(A). Then, since P(B|A') = P(B), 1 − P(B|A') = P(B'|A') = 1 − P(B) = P(B').

4.37 No. They are dependent since P(A|B) = 0 ≠ P(A).

4.39 (a) If there is randomness, and all children have an equal chance of being male or female, then there are 2x2x2x2x2=32 possible simple events, each with a 1/32 chance. Therefore, the probability of 5 boys is 1/32 and the probability of mffmf is also 1/32.

(b) P(NNNNN) = (.8)(.8)(.8)(.8)(.8) = .0352, while P(DNNDN) = (.2)(.8)(.8)(.2)(.8) = .0022. Therefore, NNNNN is more probable.

4.41 No. P(1,1) = 1/36 so fair odds would be 35:1. We expect to pay 1/36 of the time. (1/36)1000 = 27.78 times x $31 = $861.11, noting that with 30:1 odds, the bettor gets a $30 payoff, plus the return of the original $1 bet. Therefore, expected profit = $1000 − 861.11 = 138.89.

4.43 P(C_1) = .89 and P(C_2|C_1') = .75. P(not caught) = P(C_1' ∩ C_2') = P(C_1')P(C_2'|C_1') = .11(.25) = .0275

4.45 (a) (4/10)(3/9)(2/8) = 1/30 = .033

(b) P(at least 1 member from Finance) = 1 − P(no members from Finance) = 1 − (6/10)(5/9)(4/8) = 1 − 1/6 = 5/6 = .833

4.47 P(C ∩ D) = P(C)P(D) since they are independent = .56(.35) = .196

4.49 .3(.74) = .222

4.51 (a) .67(.85) = .5695

(b) P(not permitted to pay by check) = P(L_1 ∪ L_2) = .67 + .85 − (.67)(.85) = .9505

(c) P(accept check) = P(do not appear on either list) = P(L_1' ∩ L_2') = .33(.15) = .0495

4.53 (a) .7(.7) = .49 (b) .3(.3) = .09

(c) P(correct results) = .85(.85) = .7225, P(incorrect results) = .15(.15) = .0225

4.55 Let A_1 = the event that the person is hired, A_2 = the event that the person is not hired, B = the event that the person has previous experience. $P(A_1) = .31$, $P(A_2) = .69$, $P(B|A_1) = .74$, $P(B|A_2) = .29$, $P(B'|A_1) = .26$, $P(B'|A_2) = .71$

$P(A_1) = .31$

$$P(A_1|B') = \frac{P(A_1)P(B'|A_1)}{P(A_1)P(B'|A_1) + P(A_1)P(B'|A_1)} = \frac{.31(.26)}{.31(.26) + .69(.71)} = .14$$

4.57

Outcomes	$P(A_i)$	$P(B|A_i)$	Joint	Revised*
H	.25	.35	.0875	.1654
M	.35	.57	.1995	.3771
L	.10	.62	.0620	.1172
A	.30	.60	.1800	.3403
			.5290	

*To get the revised probabilities, we divided $P(B|A_i)/\Sigma P(B|A_i) = P(B|A_i)/.5290$

initial: .25, .35, .10, .30   revised: .1654, .3771, .1172, .3403

4.59 No. It is possible that some companies could do more than one of these. If potential intersections exist, then we need to know their probabilities that they resorted to none of these. If, these categories are exclusive, then we would have $P(none) = 1 - P(A \cup B \cup C) = 1 - P(A) - P(B) - P(C)$.

4.61 (a) We use the mno rule, since we have the choice of one from each group.

(b) 2 x 3 x 5 = 30

(c) S = {E_4T_aN, E_4T_aS, E_4T_aR, $E_4T_aR_b$, $E_4T_aS_b$, E_4T_4N, E_4T_4S, E_4T_4R, $E_4T_4R_b$, $E_4T_4S_b$, E_4T_5N, E_4T_5S, E_4T_5R, $E_4T_5R_b$, $E_4T_5S_b$, E_6T_aN, E_6T_aS, E_6T_aR, $E_6T_aR_b$, $E_6T_aS_b$, E_6T_4N, E_6T_4S, E_6T_4R, $E_6T_4R_b$, $E_6T_4S_b$, E_6T_5N, E_6T_5S, E_6T_5R, $E_6T_5R_b$, $E_6T_5S_b$}

(d) With automatic transmission, there are 2 choices for the engine and 5 choices for color, or 2x5 = 10 total choices.

4.63 Let D be the event that interest rates decline. Let I be the event that home values increase

(a) $P(D \cap I) = P(D)P(I|D) = .6(.75) = .45$

(b) $P(I) = P(D \cap I) + P(D' \cap I) = .45$ (from before) $+ .4(.2) = .53$

(c) $P(D'|I) = \frac{P(D' \cap I)}{P(I)} = \frac{.4(.2)}{.53} = .151$

4.65 $1 - (.68 + .20 + .05) = .07$

4.67 (a) $P(C|P) = \frac{P(C \cap P)}{P(P)} = \frac{10/200}{14/200} = .714$

(b) $(26+28+12)/200 = .33$

(c) $P(A \cap P) = 2/200 = .01$

(d) $P(G|B) = \frac{20/200}{50/200} = .4$

4.69 mutually exclusive: c, d, e   not mutually exclusive: a, b, f (you can own more than one policy)

4.71 $P(C \cap L) = P(C)P(L|C) = .8(.3) = .24$

4.73 (a) $(\frac{20}{50})(\frac{19}{49}) = .1551$ (b) $(\frac{30}{50})(\frac{29}{49}) = .3551$

(c) P(one man and one woman) = $P(M_1 \cap W_2) + P(W_1 \cap M_2) = (\frac{20}{50})(\frac{30}{49}) + (\frac{30}{50})(\frac{20}{49}) = 24/49 = .4898$.

4.75 $P(D) = .48$, $P(C) = .23$, and $P(D \cup C) = .55$. Using the additive rule and solving for $P(D \cap C)$, we have $P(D \cap C) = .48 + .23 - .55 = .16$.

4.77 1/24

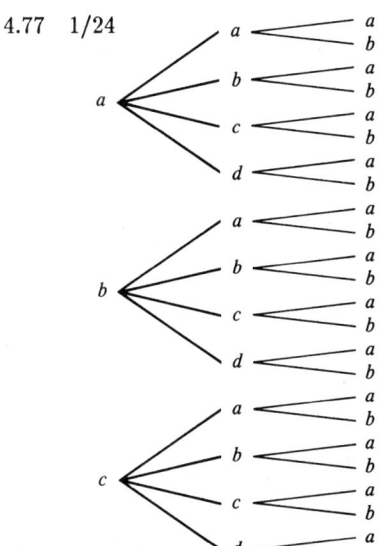

4.79

(a) .76 (b) .53 (c) .85

4.81

Outcome	$P(A_i)$	$P(B|A_i)$	Joint	Revised
Profitable	.45	.80	.36	.36/.47 = .7660
Not	.55	.20	.11	.11/.47 = .2340
			.47	

P(Profitable|Aggressive Campaign) = .7660

4.83 $\binom{8}{5} = \frac{8!}{5!3!} = 56$

4.85 (a) P(FC|Female) = $\frac{.50}{.55} = .909$ (b) P(FC $\cap$ Female) = .50

(c) P(OC | Male) = 15/45 = 1/3 (d) P(FC) = .8

(e) P(OC | Male) = 1/3 from (c) P(OC) = .2. Since these are not equal, then the position is not independent of sex.

4.87 Using the information given in the problem, we now construct the table below:

	Sunbelt	Elsewhere	Totals
Growth	80	38	118
No Growth	137	95	232
Totals	217	133	350

(a) $P(G|S) = \dfrac{80/350}{217/350} = 0.37$ (b) P(S) = 217/350 = .62
(c) P(G') = 232/350 = .663 (d) P(S∩G) + P(S'∩G') = (80+95)/350 = .5

4.89 (a) .22 (b) .22+.15 = .37 (c) .15+.11 = .26 (d) .22+.15+.11 = .48 (e) 0(these are mutually exclusive)

4.91 For two passengers, the sample space is S = {AA, AB, BA, BB}, similar to tossing two coins. With an equal chance of flying either airline, the probability that they flew on the same airline is P(AA ∪ BB) = 2/4 = .5. For three passengers, we have(similar to tossing three coins) S = {AAA, AAB, ABA, ABB, BAA, BAB, BBA, BBB). Therefore, the probability that all three flew the same airline is P(AAA) + P(BBB) = 2/8 = .25.

4.93 Either/or means union. P(O ∪ Und) = .08 + .18 = .26 (They are mutually exclusive and can be over and underfilled at the same time.

4.95 P(at least one bad) = 1 − P(no bad ones) = 1 − $(\frac{4}{6})(\frac{3}{5})$ = .6

4.97 (a) $\binom{11}{4} = \frac{11!}{4!5!} = 330$ (b) $\binom{5}{2} = 10$ (c) $\binom{6}{2} = 15$ (d) $\binom{5}{2} \times \binom{6}{2} = 150$

CHAPTER 5

PROBABILITY DISTRIBUTIONS

5.1 (a) $E[X] = 14(.09) + 15(.13) + ... + 20(.01) = 16.74$
$E[X^2] = 14^2(.09) + 15^2(.13) + ... + 20^2(.01) = 282.24$
$E(X - \mu)^2] = (14 - 16.74)^2(.09) + (15 - 16.74)^2(.13) + ... + (20 - 16.74)^2(.01) = 2.0124$
(b) $\mu = E[X] = 16.74$, $\sigma^2 = E[X - \mu]^2 = 2.0124$
(c) $\mu \pm 2\sigma$ is $16.74 \pm 2\sqrt{2.0124}$, or from 13.90 to 19.58. Therefore, P(X is within $\mu \pm 2\sigma$) = P(X = 14,15,16,17,18,19) = 1 − P(X=20) = 1 − .01 = .99

5.3 Let X = the number of men hire(d) P(X=0) = P(2 women) = (2/5)(1/4) = 1/10, P(X=2) = P(2 men) = (3/5)(2/4) = 3/10, and P(X=1) = P(1 man, 1 woman) = P(MW) + P(WM) = (3/5)(2/4) + (2/5)(3/4) = 6/10. Therefore, we have

X	p(X)
0	.1
1	.6
2	.3

$\mu = 0(.1) + 1(.6) + 2(.3) = 1.2$
$E(X^2) = 0^2(.1) + 1^2(.6) + 2^2(.3) = 1.8$
$\sigma^2 = 1.8 - 1.2^2 = .36$

5.5 (a) $\mu = 0(.38) + 1(.42) + ... + 5(.01) = 1.01$
$E(X^2) = 0^2(.38) + 1^2(.42) + ... + 5^2(.01) = 2.33$
$\sigma = \sqrt{2.33 - 1.01^2} = 1.14$
(b) $\mu \pm 2\sigma$ is $1.01 \pm 2(1.14)$, or from −1.27 to 3.29, covering X = 0,1,2,3.
$P(X = 0,1,2,3) = .38+.42+.08+.06 = .94$

5.7 (a) $\dfrac{\binom{4}{0}\binom{6}{4}}{\binom{10}{4}} = 15/210 = .07$ (b) $\dfrac{\binom{4}{1}\binom{6}{3} + \binom{4}{0}\binom{6}{4}}{\binom{10}{4}} = 95/210 = .45$

(c) $\dfrac{\binom{4}{4}\binom{6}{0}}{\binom{10}{4}} = 1/210 = .005$

5.9 (a) $P(X) = \dfrac{\binom{4}{X}\binom{2}{2-X}}{\binom{6}{2}}$ for X = 0,1,2, yielding 1/15, 8/15, and 6/15 for p(0), p(1), p(2).

− 27 −

(b) $P(X) = \dfrac{\binom{4}{X}\binom{2}{3-X}}{\binom{6}{3}}$ for X = 1,2,3 yielding 4/20, 12/20, and 4/20 for p(1), p(2), p(3).

5.11

X	p(X)
0	$\binom{6}{0}(.9)^0(.1)^6 = .000001$
1	$\binom{6}{1}(.9)^1(.1)^5 = .000054$
2	$\binom{6}{2}(.9)^2(.1)^4 = .001215$
3	$\binom{6}{3}(.9)^3(.1)^3 = .014580$
4	$\binom{6}{4}(.9)^4(.1)^2 = .098415$
5	$\binom{6}{5}(.9)^5(.1)^1 = .354294$
6	$\binom{6}{6}(.9)^6(.1)^0 = .531441$

$\mu = 0(.000001) + 1(.000054) + \ldots + 6(.531441) = np = 6(.9) = 5.4$

$E[X^2] = 0^2(.000001) + 1^2(.000054) + \ldots + 6^2(.531441) = 29.70$

$\sigma^2 = 29.70 - (5.4)^2 = npq = 6(.9)(.1) = .54$

5.13 (a)

X	p(X)
0	.6
1	.4

(b) $\mu = p = .4$, $\sigma = \sqrt{pq} = \sqrt{.4(.6)} = .4899$

(c)

Y	p(Y)
0	$\binom{5}{0}.4^0.6^5 = .078$
1	$\binom{5}{1}.4^1.6^4 = .259$
2	$\binom{5}{2}.4^2.6^3 = .346$
3	$\binom{5}{3}.4^3.6^2 = .230$
4	$\binom{5}{4}.4^1.6^4 = .077$
5	$\binom{5}{5}.4^5.6^0 = .010$

(d) $\mu = np = 5(.4) = 2$. $\sigma = \sqrt{npq} = \sqrt{5(.4)(.6)} = 1.10$

5.15 $P(X=10) = \binom{22}{10}(.65)^{10}(.35)^{12} = .0294$

assumptions: independence, randomness

5.17 $P(X<3) = P(X=0,1,2) = \binom{10}{0}.099^0.901^{10} + \binom{10}{1}.099^1.901^9 + \binom{10}{2}.099^2.901^8 = .9315$

5.19 (a) $P(X=8) = \binom{10}{8}(.7)^8(.3)^2 = .2335$ (b) $P(X=10) = \binom{10}{10}(.7)^{10}(.3)^0 = .0282$

(c) Mutually exclusive, exhaustive, independent outcomes. Independence may be lacking for the first 10 days in December, due to potential weather "streaks."

5.21 $n = 20$, $p = .05$, $\mu = 20(.05) = 1$, $P(X<1) = P(X=0) = \binom{20}{0}.05^0.95^{20} = .3585$

5.23 (a) $100{,}000(p) = 600 \Rightarrow p = 600/100{,}000 = .006$

(b) $E(\text{Payoff}) = 100{,}000(.02) = 200$.

5.25 Today, n=4, p=.2, P(X=0) = $\binom{4}{0}.2^0.8^4 = .4096$

Yesterday, n=6, P(X $\leq$ 2) = $\binom{6}{0}.2^0.8^6 + \binom{6}{1}.2^1.8^5 + \binom{6}{2}.2^2.8^4 = .9011$

P(0 today and no more than 2 yesterday) = (.4096)(.9011) = .3691

P(at least 2 reds from last two days[n=9]) = 1 − [P(X=0)+P(X=1)] =

$1 - \binom{10}{0}.2^0.8^{10} - \binom{10}{1}.2^1.8^9 = .6242$

5.27 Answers will vary here. If you make binomial tables for n=100 and p=.15, you will find a variety of intervals for a and b such that P(a $\leq$ X $\leq$ b) is at least .8. Some possibilities are 0 − 18, 12 − 100, 11 − 20, and 10 − 19. Others exist as well.

5.29 (b) To find the expected frequencies for each value, we will multiply 50 times the probability of the value. For P(X=0), if you use the pdf command for MINITAB, you get .0003, so the expected frequency is 50(.0003) = .015. As P(X=1) = .0027, the expected frequency is 50(.0027) = .135. Similarly, for the other values, the expected values will be 50P(X=k), yielding .535, 1.43, 2.865, 4.58, 6.105, 6.98, 6.205, 4.965, 3.61, 2.405, 1.8, .845, .45, .225, and .105 for the values 2 through 20 respectively.

5.31 (a) P(X=0) = $\frac{e^{-4} 4^0}{0!} = .0183$ (b) P(X=1) = $\frac{e^{-4} 4^1}{1!} = .0733$

(c) P(X=3) = $\frac{e^{-4} 4^3}{3!} = .1954$ (d) P(X=5) = $\frac{e^{-4} 4^5}{5!} = .1563$

(e) P(X $\geq$ 2) = 1 − P(X $\leq$ 1) = 1 − (.0183+.0733) = .9084

5.33 10 years = 120 months. λ = 216/120 = 1.8

P(X $\geq$ 2) = 1 − P(X=0,1) = 1 − $e^{-1.8} 1.8^0/0!$ − $e^{-1.8} 1.8^1/1!$ = .5372

5.35 λ = (1320/30)(1/4)(1/3) = 3.67 note the 1/4 represents 6 hours in a 24 hour day and 1/3 represents the one of three members per shift.

(a) P(X=0) = $\frac{e^{-3.67} 3.67^0}{0!} = .0255$ (b) P(X=1) = $\frac{e^{-3.67} 3.67^1}{1!} = .0935$

(c) P(X $\geq$ 1) = 1 − P(X=0) = 1 − .0255 = .9745

(d) P(X $\geq$ 2) = 1 − P(X=0,1) = 1 − (.0255+.0935) = .8810

5.37 Poisson: λ = 30(1/50) = .6 P(X=0) = $e^{-.6}(.6)^0/(0!) = .5488$

Binomial: P(X=0) = $\binom{30}{0}(.02)^0(.98)^{30} = .5455$

5.39 (a) If P(X=1)=P(X=2), then we have $\frac{\lambda^1 e^{-\lambda}}{1!} = \frac{\lambda^2 e^{-\lambda}}{2!}$, so $\lambda=2=\mu=\sigma^2$. Therefore, μ=2 and $\sigma=\sqrt{2}$.

(b) P(X=0) = $\frac{2^0 e^{-2}}{0!} = e^{-2} = .1353$

(c) P(X=1) = $\frac{2^1 e^{-2}}{1!} = .2707$

(d) P(X=2) = $\frac{2^2 e^{-2}}{2!} = .2707$

(e) P(1 or 2) = P(X=1)+P(X=2) = .2707 + .2707 = .5414

5.41 To find these probabilities, it may be helpful to draw a picture. When finding the area between two points, if both points are positive, find the central area to each point and subtract the smaller from the larger, and the same for when both points are negative. When one point is positive and one negative, find the central area to each point and add the areas.

(a) 41.92% (b) 49.38% (c) 48.61 − 43.32 = 5.29% (d) 22.57 − 19.15 = 3.42%
(e) 41.92+31.59 = 73.51% (f) 47.88 − 46.93 = .95% (g) 43.82 − 41.31 = 2.51%
(h) 48.38+40.82 = 89.20% (i) 50% (j) 49.51+46.49 = 96% (k) 49.87 − 44.74 = 5.13%
(l) 42.92 − 35.54 = 7.38% (m) 49.27+17.36 = 66.63% (n) 16.64+43.06 = 59.70%
(o) 11.03 − 6.75 = 4.28%

5.43 (a) ±0.67 (b) 1.65 (c) 1.28 (d) ±.67 (e) ±.23 (f) ±1.96 (g) ±2.58

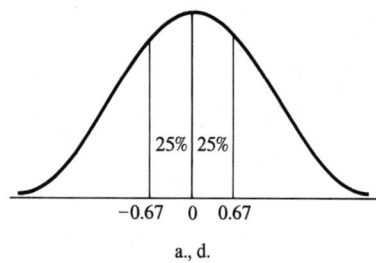

a., d.

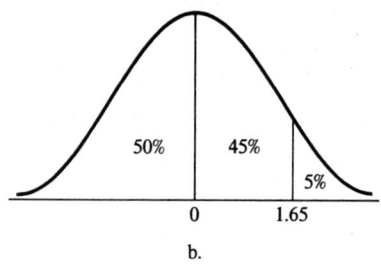

b.

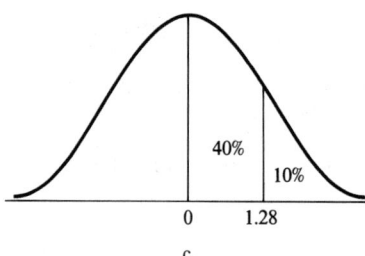

c.

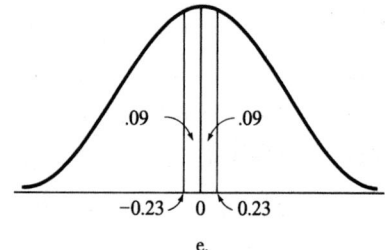

e.

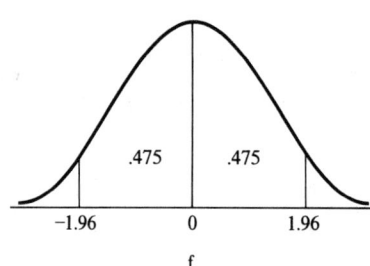

f.

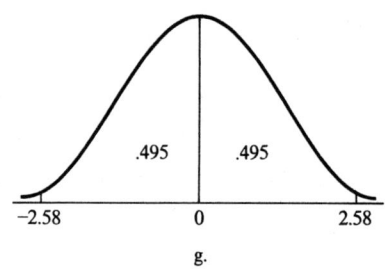

g.

5.45 (a) $z = 0.9$ (b) $z = -1.47$ (c) $z_a = -0.61$, $z_b = 0.09$

(d) $z_a = -1.34$, $z_b = 0.97$

(e) The area from 0 to 1.78 is .4625. Therefore, the area from 0 to z is .4625 − .2203 = .2422, and $z = 0.65$.

(f) The area from 0 to 0.71 is .2611, so the area from 0 to z is .2611 + .2102 = .4713, so $z = 1.90$.

(g) The area from -0.65 to 0 is .2422, so the area from 0 to $z = .3830$, so $z = -1.19$.

(h) The area from -3 to 0 is .4987, so the area from 0 to $z = .4987 - .2730 = .2257$, so $z = -0.60$

5.47 (a) $z = -1.65$, $X = \mu - 1.65\sigma = 125 - 1.65(30) = 75.5$

(b) $z = -0.92$, $X = 125 - 0.92(30) = 97.4$

(c) $X = 125 + 1.73(30) = 176.9$

(d) When you draw the picture, z will be greater than .65 of the curve, so that the area from 0 to z is .15 => $z = 0.39$. Therefore, $X = 125 + 0.39(30) = 136.7$.

(e) $z = (140 - 125)/30 = .5$. $P(z<.5) = .5 + .1915 = .6915$, so $PR_{140} = 69$.

(f) $z = (50 - 125)/30 = -2.5$. $P(z<-2.5) = .5 - .4938 = .0062$, so $PR_{50} = 1$.

(g) With central area $= .45/2 = .225$ on each side, $z = 0.60$. Therefore, the interval is from $\mu - 0.6\sigma$ to $\mu + 0.6\sigma$, or from $125 - 18 = 107$ to $125 + 18 = 143$.

5.49 It will help here to draw a picture. If the area is less than 50%, then the z-score will be negative and you will find it by looking up the area of 50% − (a) If A is more than 50%, then the z-score will be positive, and you will look up the area of A − 50%. Then the term we are looking for will be $\mu + z\sigma$ if z is positive, and $\mu - z\sigma$ if the location is below the mean.

(a) $X = 100 - 1.13(10) = 88.7$ (b) $X = 100 - 0.52(10) = 94.8$

(c) $z = 0$, $X = 100$ (d) $X = 100 + 0.67(10) = 106.7$

(e) $X = 100 + 0.84(10) = 108.4$ (f) $X = 100 - 0.95(10) = 90.5$

(g) $X = 100 - 0.52(10) = 105.2$ (h) $X = 100 + 1.28(10) = 112.8$

(i) $X = 100 - 0.39(10) = 96.1$ (j) $X = 100 + 0.23(10) = 102.3$

5.51 (a) $z = -1.98$ (b) The area from 0 to z is $.2673 + .2125 = .4798$, so $z = 2.05$.

(c) The area from 0 to z is $.4678 - .1240 = .3438$, so $z = -1.01$.

(d) The area from 0 to z is $.7959 - .4474 = .3485$, so $z = -1.03$.

(e) $z_1 = (36.5 - 35)/1 = 1.5$, so the area from 0 to $X = .6386 - .4332 = .2054$. $X = \mu - z\sigma$ where $z = 0.54$, so $X = 35 - 0.54(1) = 34.46$.

(f) X=124, so $z = (124 - 200)/50 = -1.52$. The area form 0 to X is $.4357 - .2549 = .1808$, so $z = -0.47$, and $X = 200 - 0.47(50) = 176.5$.

(g) X=53.8, $z = (53.8 - 50)/10 = 0.38$. The area from 0 to X is $.1480 + .3332 = .4812$, so $z = 2.08$ and $X = 50 + 2.08(10) = 70.8$.

(h) The area from 0 to $X = .5 - .3974 = .1026$, so $z = 0.26$, so $z = 350 - 0.26(25) = 343.50$.

5.53 (a) $P(.3400<X<.3500) = P(\frac{.3400 - .3456}{.0078} < z < \frac{.3500 - .3456}{.0078}) = P(-0.72<z<0.56) = .2642 + .2123 = .4765$, or 47.65%.

(b) $P(X>.3560) = P(z>\frac{.3560 - .3456}{.0078}) = P(z>1.33) = .5 - .4092 = .0918$ or 9.18%.

(c) $P(X<.3232) = P(z<\frac{.3232 - .3456}{.0078}) = P(z<-2.87) = .5 - .4979 = .0021$, or 0.21%

5.55 (a) $P(14<X<18) = P(\frac{14-15}{4}<z<\frac{18-15}{4}) = P(-0.25<z<0.75) = .0987+.2734 = .3721$, or 37.21%.

(b) $P(X>20) = P(z>\frac{20-15}{4}) = P(z>1.25) = .5 - .3944 = .1056$. $200(.1056) = 21.12$ or ≈ 21.

(c) Then, we have 5% in the right tail and 45% in the middle. Therefore, $z = 1.65$ and $X = \mu+1.65\sigma = 15+1.65(4) = 21.6$ mph over the speed limit, or $45+21.6=66.6$ mph.

5.57 Here, it is best to answer part c first, since we know that $P(X>200) = 165/N$.

$P(X>200) = P(z>\frac{200-100}{50}) = P(z>2) = .5 - .4772 = .0228 = 165/N$, so $N = 165/.0228 = 7236.84$, or 7237.

(a) $P(170<X<210) = P(\frac{170-100}{50}<z<\frac{210-100}{50}) = P(1.4<z<2.2) = .4861 - .4192 = .0669$. The expected number in this interval is $.0669(7237) = 484.16$ or 484.

(b) $P(X<0) = P(z<\frac{0-100}{50}) = P(z<-2) = .5 - .4772 = .0228$, so the expected number is $.0228(7237) = 165$.

5.59 $P(56,000<X<57,000) = P(\frac{56,000-55,500}{4000}<z<\frac{57,000-55,500}{4000}) = P(.13<z<.37) = .1443 - .0517 = .0926 = 215/N$, so $N = 215/.0926 = 2321.81$ or 2322.

$P(X<49,000) = P(z<\frac{49,000-55,500}{4000}) = P(z<-1.63) = .5 - .4484 = .0516$, so the expected number is $.0516(2322) = 119.82$ or 120.

5.61 Assuming symmetry, $\mu = 28$. $28+1.96\sigma = 31$, so $\sigma = 3/1.96 = 1.53$.

$P(X>30) = P(z>\frac{30-28}{1.53}) = P(z>1.31) = .5 - .4049 = .0951$

Let X = the number of cars obtaining more than 30 mpg. Then X is a binomial variable with n=5 and p = .0951. $P(X=1) = \binom{5}{1}.0951^1.9049^4 = .3188$

5.63 (a) Since we need to concentrate only on those clients with less than a year's service by the competitor, first we find $P(X<1)$. With $\sigma=8$ months, we convert this to $8/12 = 2/3$ years, to be consistent with the mean.

$P(X<1) = P(z<\frac{1-2}{2/3}) = P(z<-1.5) = .5-.4332 = .0668$. Therefore, we expect $.0668(80) = 5.34$ of the companies to have its present system for less than a year. With a 40% chance of changing, we expect $5.34(.40) = 2.136$ companies to change each week.

(b) If we let X=the number of companies that have had its present system for less than a year. Then X is a binomial variable with n=80 and p=.0668. Then,

$P(X=1) = \binom{80}{1}.0668^1.9332^{79} = .0227$

5.65 With $P(X \geq 10) = .25$, then the value for 10 is in the upper 25% of the curve. With 25% in the middle, this generates a z-score of 0.67, so we have $(10 - \mu)/\sigma = 0.67$, or $\mu + 0.67\sigma = 10$. For $P(X \leq 8) = .40$, this is in the lower part of the curve with .10 of the area in the middle, yielding a z-score of -0.25. This leaves us with two equations and two unknowns.

$\mu + 0.67\sigma = 10$

$\underline{\mu - 0.25\sigma = 8}$

$0.92\sigma = 2$, so $\sigma = 2/0.92 = 2.17$, and $\mu = 8 + 0.25(2.17) = 8.54$

5.67 $\lambda = 12$. (a) $P(X>1/6) = e^{-12/6} = e^{-2} = .1353$

(b) $P(X<1/12)P(X>1/12) = (1-e^{-1})e^{-1} = .6321(.3679) = .2325$

(c) $\mu = \sigma = 1/12$. $\mu \pm 2\sigma$ goes from 0 to 3/12. $P(X<1/4) = 1 - e^{-12/4} = 1 - e^{-3} = 1 - .0498 = .9502$.

5.69 $\mu = np = 35(.03) = 1.05$. $P(X=0) = e^{-1.05}(1.05)^0/(0!) = .3499$

5.71 (a) $P(83{,}600<X<85{,}200) = P(\frac{83{,}600 - 79{,}500}{8000} < z < \frac{85{,}200 - 79{,}500}{8000}) = P(.51<z<.71) = .2611 - .1950 = .0661 = 30/N$, so $N = 453.86$ or 454.

(b) $P(X<65{,}000) = P(z<\frac{65{,}000 - 79{,}500}{8000}) = P(z<-1.81) = .5 - .4649 = .0351$. $.0351(454) = 15.94$ or 16.

(c) $P(65{,}000<X<85{,}000) = P(\frac{65{,}000 - 79{,}500}{8000} < z < \frac{85{,}000 - 79{,}500}{8000}) = P(-1.81<z<0.69) = .4649 + .2549) = .7198$. $.7198(454) = 326.69$ or 327.

5.73 For this problem, we suggest drawing a picture. For $P(z<a)$, if a is positive, the area is $.5 +$ the area from 0 to a in the table. If a is negative, the area is $.5 -$ the area from 0 to a in the table.

(a) $P(z<1.00) = .5 + .3413 = .8413$ (b) $P(z<1.50) = .5 + .4332 = .9332$

(c) $P(z<-2.00) = .5 - .4772 = .0228$ (d) $P(z<-2.50) = .5 - .4938 = .0062$

(e) $P(z<1.20) = .5 + .3849 = .8849$ (f) $P(z<1.23) = .5 + .3907 = .8907$

(g) $P(z<-3.04) = .5 - .4988 = .0012$ (h) $P(z<.87) = .5 + .3078 = .8078$

(i) $P(z<-.94) = .5 - .3264 = .1736$ (j) $P(z<.07) = .5 + .0279 = .5279$

5.75 $\mu = 120$, $\sigma = 10$, (a) $P(X \geq 125) = P(z>\frac{125-120}{10}) = P(z>.5) = .5 - .1915 = .3085$.

(b) $P(X<T) = .85$, $z = 1.04$, $T = \mu + 1.04\sigma = 120 + 1.04(10) = 130.4$

(c) $P(X>132) = P(z>\frac{132-120}{10}) = P(z>1.2) = .5 - .3849 = .1151$

5.77 (a) $P(X=7) = \binom{12}{7}(.6)^7(.4)^5 = .2270$ (b) .2270

(c) mutually exclusive, exhaustive, independent Bernoulli trials. Independence may be lacking for student council members.

5.79 $P(X=25) = e^{-20}(20)^{25}/(25!) = .0446$

5.81   .9616 − 1.0224

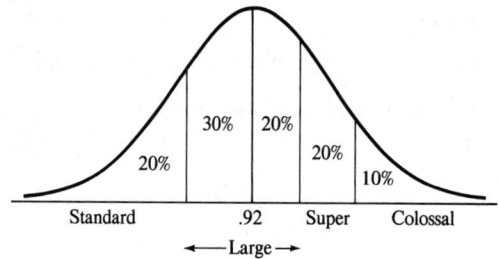

5.83 (a) z = 0, PR = 50   (b) z = (85 − 90)/20 = − .25, PR = .5 − .0987 = .4013, or 40

(c) z = (103 − 90)/20 = .65, PR = .5+.2422 = .7422, or 74

(d) z = (62 − 90)/20 = − 1.4, PR = .5 − .4192 = .0808, or 8

(e) z = (55 − 90)/20 = − 1.75, PR = .5 − .4599 = .0401, or 4

(f) z = (115 − 90)/20 = 1.25, PR = .5+.3944 = .8944, or 89

(g) z = (99.8 − 90)/20 = .49, PR = .5+.1879 = .6879, or 69

(h) z = (40.3 − 90)/20 = − 2.49, PR = .5 − .4936 = .0064, or 1

(i) z = (77.5 − 90)/20 = − .63, PR = .5 − .2357 = .2643, or 26

(j) z = (105.7 − 90)/20 = .79, PR = .5+.2852 = .7852, or 79

5.85 (a) 30 calls per hour = 30/12 = 2.5 calls per five minutes(1/12 hour).

$P(X=1) = e^{-2.5} 2.5^1/1! = .2052$

(b) $\lambda = 30/20 = 1.5$. $P(X=0) = e^{-1.5} 1.5^0/0! = .2231$

5.87 Both have terms reflecting number of successes. The binomial distribution is for n(fixed) Bernoulli trials, while the Poisson distribution is for time intervals or space continuum.

5.89 (a) $P(13<X<14) = P(\frac{13-15}{2.5}<z<\frac{14-15}{2.5}) = P(-.8<z<-.4) = .2881 - .1554 = .1327$

(b) $P(X<12) = P(z<\frac{12-15}{2.5}) = P(z<-1.2) = .5 - .3849 = .1151$

(c) $P(X<18) = P(z<\frac{18-15}{2.5}) = P(z<1.2) = .5+.3849 = .8849$

(d) $P(12<X<14.5) = P(\frac{12-15}{2.5}<z<\frac{14.5-15}{2.5}) = P(-1.2<z<-.2) = .3849 - .0793 = .3056 = 500/N$. N = 500/.3056 = 1636.13 or 1636.

5.91

X	p(X)
0	$\binom{5}{0}.22^0.78^5 = .2887$
1	$\binom{5}{1}.22^1.78^4 = .4072$
2	$\binom{5}{2}.22^2.78^3 = .2297$
3	$\binom{5}{3}.22^3.78^2 = .0648$
4	$\binom{5}{4}.22^1.78^4 = .0091$
5	$\binom{5}{5}.22^5.78^0 = .0005$

$\mu = np = 5(.22) = 1.1, \sigma = \sqrt{npq} = \sqrt{5(.22)(.78)} = .9263$

5.93 To find the probability of X successes on n Bernoulli trials, those with 2 mutually exclusive, exhaustive, independent outcomes.

5.95 To find probabilities where we are sampling without replacement from a finite group with known original conditions(numbers).

5.97 $\lambda = .5(2.4) = 12$ (a) $P(X=10) = e^{-12} 12^{10}/(10!) = .1048$

(b) with $\lambda = 2.4$, $P(X=0) = e^{-2.4} 2.4^0/(0!) = .0907$

(c) $P(X<.5) = 1 - e^{-2.4(.5)} = 1 - e^{-1.2} = 1 - .3012 = .6988$

5.99 (a) $P(X<65) = P(z<\frac{65-92}{23}) = P(z<-1.17) = .5 - .3790 = .1210$

(b) $P(X>77) = P(z>\frac{77-92}{23}) = P(z>-0.65) = .5+.2422 = .7422$

(c) $P(X<112) = P(z<\frac{112-92}{23}) = P(z<-0.87) = .5+.3078 = .8078$

(d) $P(X>120) = P(z>\frac{120-92}{23}) = P(z>1.22) = 5 - .3888 = .1112$

(e) $P(69<X<115) = P(\frac{69-92}{23}<z<\frac{115-92}{23}) = P(-1<z<1) = 2(.3413) = .6826$

(f) $P(92<X<100) = P(0<z<\frac{100-92}{23}) = P(0<z<0.35) = .1368$

(g) $P(100<X<108) = P(\frac{100-92}{23}<z<\frac{108-92}{23}) = P(0.35<z<0.70) = .2580 - .1368 = .1212$

5.101 (a) .5 (b) .2580 (c) .4906 (d) .4554 − .4452 = .0102 (e) .4893 − .1554 = .3339

(f) .4987 − .4474 = .0513 (g) .4292 − .3554 = .0738 (h) .4927+.1736 = .6663

(i) .1664+.4306 = .5970 (j) .1103 − .0675 = .0428

5.103

X	p(X)
5.5	.04
15.5	.18
25.5	.26
35.5	.21
45.5	.25
55.5	.06

(a) $\mu = 5.5(.04) + 15.5(.8) + ... + 55.5(.06) = 31.8$

$E(X^2) = 5.5^2(.04) + 15.5^2(.8) + ... + 55.5^2(.06) = 1180.55$

$\sigma = \sqrt{1180.55 - (31.8)^2} = 13.01$

(b) 31.8   (c) 31.8 x $18.75 = $596.25   (d) 596.25 − 525 = $71.25

5.105 (a) $P(X \geq 79.5) = .17$ With .17 in the upper tail, there is $.5 - .17 = .33$ in the middle, so $z = 0.95$, and we write $74.8 + 0.95\sigma = 79.5$. Therefore, $\sigma = 4.7/.95 = 4.95$.

(b) Here, we have $P(X<65) = .2$, with $\sigma = 15$. With $z = -0.84$, we write $\mu - 0.84(15) = 65$, so $\mu = 65 + 12.6 = 77.6$

5.107 (a) $\dfrac{\binom{12}{1}\binom{8}{1}\binom{4}{1}}{\binom{24}{3}} = \dfrac{384}{2024} = .1897$

(b) P(all three from same data) = P(all three from 7 day) + P(3 from 9 days) +

$P(3 \text{ from } 11 \text{ days}) = \dfrac{\binom{12}{3} + \binom{8}{3} + \binom{4}{3}}{\binom{24}{3}} = \dfrac{220+56+4}{2024} = .1383$

5.109 (a) $P(X<21) = P(z < \dfrac{21-22}{2}) = P(z<-0.5) = .5 - .1915 = .3085$

(b) $P(20 \leq X \leq 25) = P(\dfrac{20-22}{2} \leq z \leq \dfrac{25-22}{2}) = P(-1 \leq z \leq 1.5) = .3413+.4332 = .7745$. $n = .7745(48) = 37.18$ or 37.

(c) We need to find D such that $P(X>D) = .10$ (only 10% of the cases will be longer than D), so the z-score for $.5 - .1 = .4$ is 1.28 and $D = 22+1.28(2) = 24.56$

5.101 No. It only takes 1.8 standard deviations below the mean to get down to 0. We would not have an apartment with 0 square feet, but for the "minimum" size apartment, we would violate the symmetry of the distribution. If the smallest apartment was, for example, 500 square feet, this would yield a z-score of $(500 - 900)/500 = -0.8$, and $P(z<-0.8) = .5 - .2881 = .2119$, not 0.

5.111 No. It only takes 1.8 standard deviations below the mean to get down to 0. We would not have an apartment with 0 square feet, but for a minimum size apartment, we would violate the symmetry of the distribution. If the smallest apartment was 500 square feet, this would yield a z-score of $(500 - 900)/500 = -0.8$, and if the distribution was normal, $P(z < -0.8) = .5 - .2881 = .2119$, not 0.

CHAPTER 6

SAMPLING DISTRIBUTIONS AND SAMPLING PROCEDURES

6.1 (a) $\Sigma X = 55$, $\Sigma X^2 = 4+49+81+196+529 = 859$, $\mu = 55/5 = 11$, $\sigma^2 = \dfrac{859 - \dfrac{(55)^2}{5}}{4} = 50.8$, $\sigma = \sqrt{50.8} = 7.13$

(b) (2,7)(2,9)(2,14)(2,23)(7,9)(7,14)(7,23)(9,14)(9,23)(14,23)

(c) 4.5,5.5,8,12.5,8,10.5,15,11.5,16,18.5

(d) $\Sigma \overline{X} = 110$, $\Sigma \overline{X}^2 = 1400.50$, $\mu_{\overline{X}} = 110/10 = 11$, $\sigma_{\overline{X}}^2 = \dfrac{1400.50 - \dfrac{(110)^2}{10}}{10} = 19.05$, $\sigma = \sqrt{19.05} = 4.36$. We should have $\sigma_{\overline{X}} = \dfrac{\sigma}{\sqrt{n}}\sqrt{\dfrac{N-n}{N-1}} = \dfrac{7.13}{\sqrt{2}}\sqrt{\dfrac{5-2}{5-1}} = 4.37$

6.3 (a) $\Sigma X = 59$, $\Sigma X^2 = 727$, $\mu = 59/6 = 9.83$, $\sigma^2 = \dfrac{727 - (59)^2/6}{6} = 24.47$, $\sigma = \sqrt{24.47} = 4.95$

(b) (8,6)(8,5)(8,9)(8,11)(8,20)(6,5)(6,9)(6,11)(6,20)(5,9)(5,11)(5,20)(9,11)(9,20)(11,20)

(c) 7,6.5,8.5,9.5,14,5.5,7.5,8.5,13,7,8,12.5,10,14.5,15.5

(d) $\Sigma \overline{X} = 147.5$, $\Sigma \overline{X}^2 = 1597.25$, $\mu_{\overline{X}} = 147.5/15 = 9.83$, $\sigma_{\overline{X}}^2 = \dfrac{1597.25 - (147.5)^2/15}{15} = 9.79$, $\sigma_{\overline{X}} = \sqrt{9.79} = 3.13$

(e) With $\mu_{\overline{X}} = \mu = 9.83$, and $\sigma_{\overline{X}} = \dfrac{\sigma}{\sqrt{n}} = \dfrac{4.95}{\sqrt{2}} = 3.50$, the results do not agree. We need to apply the finite correction factor for $\dfrac{\sigma}{\sqrt{n}}\sqrt{\dfrac{N-n}{N-1}} = \dfrac{4.95}{\sqrt{2}}\sqrt{\dfrac{6-2}{6-1}} = 3.13$ to agree with the result from part d.

6.5 (a) $P(\overline{X}>8) = P(z>\dfrac{8-7.9}{2.70/\sqrt{45}}) = P(z>.25) = .5 - .0987 = .4013$

(b) $P(\overline{X}<7) = P(z<\dfrac{7-7.9}{2.70/\sqrt{45}}) = P(z<-2.24) = .5 - .4875 = .0125$

(c) $P(7.5<\overline{X}<8.5) = P(\dfrac{7.5-7.9}{2.70/\sqrt{45}}<z<\dfrac{8.5-7.9}{2.70/\sqrt{45}}) = P(-0.99<z<1.49) = .3389+.4319 = .7708$

6.7 (a) $\mu_{\overline{X}} = 500$, $\sigma_{\overline{X}} = 25/\sqrt{16} = 6.25$

(b) $P(\overline{X}<495) = P(z<\dfrac{495-500}{6.25}) = P(z<-0.8) = .5-.2881 = .2119$, or 21.19%

(c) $P(X<495) = P(z<\dfrac{495-500}{25}) = P(z<-0.2) = .5-.0793 = .4207$

6.9 (a) $P(\overline{X} \geq 55{,}400) = P(z \geq \frac{55{,}400 - 55{,}500}{4000/\sqrt{2000}}) = P(z \geq -1.12) = .5 + .3686 = .8686$

(b) $P(55{,}450 \leq \overline{X} \leq 55{,}550) = P(\frac{55{,}450 - 55{,}500}{4000/\sqrt{2000}} \leq z \leq \frac{55{,}550 - 55{,}500}{4000/\sqrt{2000}}) = P(-0.56 \leq z \leq 0.56) = 2(.2123) = .4246$

(c) $P(X < 40{,}000) = P(z < \frac{40{,}000 - 55{,}500}{4000}) = P(z < -3.875) \approx 0$

(d) 0 (this is the probability of a point)

(e) This would leave 10% in the right side central area of the curve, yielding $z = 0.25$. Therefore, L $= \mu + 0.25\sigma = 55{,}500 + 0.25(4000) = 56{,}500$

6.11 $P(\overline{X} > 19) = P(z < \frac{19 - 18.5}{3/\sqrt{6}}) = P(z < 0.41) = .5 + .1591 = .6591$, or 65.91%

6.13 (a) $P(\overline{X} < 195{,}000) = P(z < \frac{195{,}000 - 200{,}000}{25{,}000/\sqrt{90}}) = P(z < -1.90) = .5 - .4713 = .0287$

(b) $P(\overline{X} \geq Q) = .75$, then the area from the mean to our quota, Q, is .25, yielding z $= -0.67$, so $Q = 200{,}000 - 0.67(25{,}000/\sqrt{90}) = 198{,}234$ barrels per day.

6.15 (a) No. There is no balance or symmetry. There is a high percentage of homes in the lower left categories.

(b) $P(\overline{X} \geq 30{,}000) = P(z \geq \frac{30{,}000 - 29475}{23{,}739/\sqrt{400}} = P(z \geq 0.44) = .5 - 1700 = .3300$

(c) No. We are finding a probability about $\overline{X}$, the sample mean, not the individual incomes. With n=2000, the Central Limit Theorem says that the sampling distribution of $\overline{X}$ is approximately normal.

6.19 $P(\Sigma X > \$144{,}000) = P(z > \frac{144{,}000 - 36(4200)}{\sqrt{36}(700)} = P(z > -1.71) = .5 + .4564 = .9564$

6.21 (a) yes, assembly line (b) no, this is a discrete variable (c) yes, this is an average of n>30 (d) yes, this is a sum of n>30 (e) no, the distribution is probably skewed with the mean being less than the median (f) yes, this is a sampling distribution of means (g) no, not enough information to make a determination

6.23 $P(\Sigma X \leq 280) = P(z \leq \frac{280 - 8(35.85)}{\sqrt{8}(5.85)}) = P(z \leq -0.41) = .5 - .1591 = .3409$

6.25 $P(\Sigma X > 420) = P(z > \frac{420 - 8(55)}{\sqrt{8}(10)}) = P(z > -0.71) = .5 + .2611 = .7611$

6.27 (a) $\mu_{\Sigma X} = 20(20) = 400$, $\sigma_{\Sigma X} = \sqrt{20}(1.8) = 8.05$

(b) $P(\Sigma X \leq 410) = P(z \leq \frac{410 - 400}{8.05}) = P(z \leq 1.24) = .5 + .3925 = .8925$

(c) $P(380 \leq \Sigma X \leq 420) = P(\frac{380 - 400}{8.05} \leq z \leq \frac{420 - 400}{8.05}) = P(-2.48 \leq z \leq 2.48) = 2(.4934) = .9868$

(d) $P(\Sigma X > M) = .3$. Since M is in the upper 30% of the curve, the central area is .2, giving a z-score of 0.52, so $M = 400 + 0.52(8.05) = 404.19$

6.29 $P(X \geq 79) = P(z \geq \frac{78.5 - 82}{\sqrt{100(.82)(.18)}} = P(z \geq -0.91) = .5 + .3186 = .8186$

6.31 $P(X \geq 22) = P(z \geq \frac{21.5 - 20}{\sqrt{100(.2)(.8)}}) = P(z \geq 0.38) = .5 - .1480 = .3520$

6.33 (a) $\mu = 35(.8) = 28$, $\sigma = \sqrt{35(.8)(.2)} = 2.37$
   (b) $P(X>30) = P(z>\frac{30.5 - 28}{2.37}) = P(z>1.06) = .5 - .3554 = .1446$, or 14.46%
   (c) With the lower 15% of the curve, then there is 35% in the central area, giving $z = -1.04$. Therefore, with $X - .5 = 28 - 1.04(2.37) = 25.54$, giving $X = 26$
   (d) $P(X \leq 30) = P(z \leq \frac{30.5 - 28}{2.37}) = P(z \leq 1.06) = .5 + .3554 = .8544$

6.35 $P(\hat{p}>.90) = P(z>\frac{.90 - .85}{\sqrt{(.85)(.15)/150}}) = P(z>1.71) = .5 - .4564 = .0436$

6.37 (a) Using the binomial distribution, we have $(.95)^{150} = .000456$. Using the normal approximation to the binomial, we have
   $P(X<1) = P(z<\frac{0.5 - 150(.05)}{\sqrt{150(.05)(.95)}}) = P(z<-2.62) = .5 - .4956 = .0044$
   (b) Poisson approximation, $\lambda = np = 7.5$, $P(X=0) = e^{-7.5} 7.5^0/0! = .0006$

6.39 $P(X>50) = P(z>\frac{50.5 - 60}{\sqrt{300(.2)(.8)}}) = P(X>-1.37) = .5 + .4147 = .9147$

6.41 $P(25 \leq X \leq 30) = P(\frac{24.5 - 22.5}{\sqrt{75(.3)(.7)}} \leq z \leq \frac{30.5 - 22.5}{\sqrt{75(.3)(.7)}}) = P(.50 \leq z \leq 2.02) = .4783 - .1915 = .2868$

6.43 To use the normal approximation to the binomial, we need $np \geq 5$. Therefore, $n(.35) \geq 5$, so $n \geq 5/(.35) = 14.29$, so the minimum sample size is 15.

6.45 No. There is no guarantee. Yes, it is likely.

6.47 With $\sqrt{n}$ in the denominator of the standard error terms, as n increases, the standard error term decreases.

6.49 We could either use systematic sampling and choose every 15th member, or use a random number generator and randomly select 100 members.

6.51 Systematic would make the most sense. Random is a possibility.

6.53 (a) No. We can't go 2 standard deviations below the mean to get to 0, our lowest value. Also, the number of defects is discrete.
   (b) $P(\overline{X} \leq 1.8) = P(z \leq \frac{1.8 - 2.1}{1.2/\sqrt{36}}) = P(z \leq -1.5) = .5 - .4332 = .0668$
   (c) No. The Central Limit Theorem says that the sampling distribution of $\overline{X}$ will be approximately normal.

6.55 (a) $\mu_{\overline{X}} = 50$, $\sigma_{\overline{X}} = 5/\sqrt{36} = .833$
   (b) $P(\overline{X}<49) = P(z<\frac{49 - 50}{.833}) = P(z<-1.20) = .5 - .3849 = .1151$

(c) $P(\overline{X}>47) = P(z>\frac{47-50}{.833}) = P(z>-3.60) = \approx 1$

6.57 (a) No. We can barely go one standard deviation below the mean to get down to 0.

(b) $P(\overline{X} \geq 150) = P(z \geq \frac{150-125}{120/\sqrt{36}}) = P(z \geq 1.25) = .5 - .3944 = .1056$

(c) $P(\Sigma X \geq 40,000) = P(z \geq \frac{40,000 - 250(125)}{\sqrt{250}(120)}) = P(z \geq 4.61) \approx 0$

(d) Either we had an incredible salesperson or maybe our figure for the average commission has increased since the most recent study.

6.59 $P(X \geq 20) = P(z \geq \frac{19.5 - 17.5}{\sqrt{50(.35)(.65)}}) = P(z \geq 0.59) - .5 - .2224 = .2776$

6.61 $P(\Sigma X \geq 175) = P(z \geq \frac{175 - 25(7.35)}{\sqrt{25}(1.80)}) = P(z \geq -0.97) = .5 + .3340 = .8340$

6.63 (a) $P(\overline{X} \geq 1.6) = P(z \geq \frac{1.6 - 1.5}{.5/\sqrt{100}}) = P(z \geq 2) = .5 - .4772 = .0228$, or 2.28%.

(b) $P(X \geq 1.6) = P(z \geq \frac{1.6 - 1.5}{.5}) = P(z \geq .2) = .5 - .0793 = .4207$, or 42.07%

(c) $P(1.45 \leq \overline{X} \leq 1.55) = P(\frac{1.45 - 1.5}{.5/\sqrt{100}} \leq z \leq \frac{1.55 - 1.5}{.5/\sqrt{100}}) = P(-1 \leq z \leq 1) = 2(.3413) = .6826$, or 68.26%

6.65 Systematic Advantages: Easy to use Disadvantages: Items must be numbered, or in sequence. Can't use with cyclical data

Stratified Advantages: Allows representation of various groups Disadvantages: Must be able to identify strata

Cluster Advantages: Useful when subjects are spread out geographically Disadvantages: Shouldn't be used if clusters are markedly different

6.67 (a) $P(\overline{X} \geq 13) = P(z \geq \frac{13 - 12}{2.5/\sqrt{49}}) = P(z \geq 2.8) = .5 - .4974 = .0026$

(b) $P(11.50 \leq \overline{X} \leq 12.25) = P(\frac{11.50 - 12}{2.5/\sqrt{49}} \leq z \leq \frac{12.25 - 12}{2.5/\sqrt{49}}) = P(-1.40 \leq z \leq 0.70) = .4192 + .2580 = .6772$

6.69 No. $np = 20(.15) = 3 < 5$. $P(\hat{p} \geq .1) = P(z \geq \frac{.1 - .15}{\sqrt{(.15)(.85)/40}}) = P(z \geq -0.89) = .5 + .3133 = .8133$

6.71 $P(42,900 \leq \Sigma X \leq 45,240) = P(\frac{42,900 - 52(850)}{\sqrt{52}(150)} \leq z \leq \frac{45,240 - 52(850)}{\sqrt{52}(150)}) = P(-1.20 \leq z \leq 0.96) = .3849 + .3315 = .7164$

6.73 $P(X<27) = P(z<\frac{26.5 - 40(.8)}{\sqrt{40(.8)(.2)}}) = P(z<-2.17) = .5 - .4850 = .0150$

6.75 (a) $\mu_{\overline{X}} = 10.75$, $\sigma_{\overline{X}} = 1.50/\sqrt{100} = .15$

(b) $P(\overline{X}>11) = P(z>\frac{11-10.75}{.15}) = P(z>1.67) = .5 - .4525 = .0475$

(c) $P(\Sigma X>A) = .2$, so the central area is .3, yielding $z = 0.84$, so $A = 10.75(100) + 0.84\sqrt{100}(1.5) = 1087.60$

6.77 (a) $P(\Sigma X \leq 2400) = P(z \leq \frac{2400-2470}{\sqrt{100}(2.40)}) = P(z \leq -2.92) = .5 - .4982 = .0018$

(b) $P(\Sigma X \geq 1900) = P(z \geq \frac{1900-24.70(75)}{\sqrt{75}(2.40)}) = P(z \geq 2.29) = .5 - .4890 = .0110$

6.79 Using the binomial, $P(X=0) = \binom{40}{0}.15^0.85^{40} = .0015$.

Using the normal approximation, $P(X<1) = P(z<\frac{0.5-6}{\sqrt{40(.15)(.85)}}) = P(z< -2.44) = .5 - .4927 = .0073$

6.81 $P(23 \leq X \leq 29) = P(\frac{22.5-80(.32)}{\sqrt{80(.32)(.68)}} \leq z \leq \frac{29.5-80(.32)}{\sqrt{80(.32)(.68)}}) = P(-0.74 \leq z \leq 0.93) = .2704 + .3238 = .5942$

6.83 The sample depends upon the point at which the table is entered

6.85 (a) X: parent, Y: sampling distribution of means, W: sampling distribution of sums, Z: sampling distribution of means

(b) $\mu_Y = 85,450$, $\sigma_Y = \sigma/\sqrt{45} = 2010/\sqrt{45} = 300$

$\mu_W = 45(85,450) = 3,845,250$, $\sigma_W = \sqrt{n}\sigma = \sqrt{45}(2010) = 13,483$

$\mu_Z = 85,450$, $\sigma_Z = \sigma/\sqrt{n} = 2010/\sqrt{180} = 150$

(c) Y, W, and Z. These are sampling distributions of averages and sums.

(d) X, Y, and Z all have means of 85,450

(e) False; $\sigma_Z = \sigma_Y/\sqrt{4} = \sigma_Y/2$

(f) False; $\sigma_Y = \sigma_X/\sqrt{45}$ (g) True (h) True

CHAPTER 7

ESTIMATION

7.1 $73/300 = .243$, $285/300 = .95$, $27/300 = .09$

7.3 $\Sigma X = 1374$, $\Sigma X^2 = 155{,}034$

  (a) $\overline{X} = 1374/15 = 91.6$

  (b) $s^2 = \dfrac{155{,}034 - (1374)^2/15}{14} = 2083.97$, $s = 45.65$

7.5 $\Sigma X = 9829$, $\Sigma X^2 = 11{,}173{,}891$, $\overline{X} = 9829/10 = 982.9$,

$s^2 = \dfrac{11{,}173{,}891 - (9829)^2/10}{9} = 168107.44$, $s = 410.01$

7.7 (a) $\Sigma fX = 44(4) + 45(12) + \ldots + 49(3) = 2971$,

$\Sigma fX^2 = 44^2(4) + 45^2(12) + \ldots + 49^2(3) = 2971$,

$\overline{X} = 2971/64 = 46.42$, based on 64 degrees of freedom

$s^2 = \dfrac{138021 - (2971)^2/64}{63} = 1.61$, $s = 1.27$ based on 63 degrees of freedom

  (b) These represent point estimates for μ, σ^2, and σ.

7.9 $\Sigma fX = 13125$, $\Sigma fX^2 = 1389637.5$, $\overline{X} = 13125/150 = 87.5$,

$s^2 = \dfrac{1389637.5 - (13125)^2/150}{149} = 1618.79$

7.11 (a) 1.86 (b) -1.80 (c) -2.950 (d) -2.48

7.13 The interval is $407.25 \pm z(35.89)/\sqrt{49}$. For 95%, 82%, and 75% confidence, the z-value is 1.96, 1.34, and 1.15. Therefore, the intervals are $407.25 \pm 10.05, 6.87, 5.90$

7.15 $\Sigma X = 374.7$, $\Sigma X^2 = 14758.57$, $\overline{X} = 37.47$, $s^2 = 79.84$, $s = 8.94$

A 90% interval for μ is $37.47 \pm 1.83(8.94)/\sqrt{10}$, or 37.47 ± 5.18

A 95% interval for μ is $37.47 \pm 2.26(8.94)/\sqrt{10}$, or 37.47 ± 6.39

7.17 For attitudes, we have $43.0 \pm 1.65(4.0)/\sqrt{155}$, or 43.0 ± 0.53

For teamwork, we have $4.5 \pm 1.65(0.98)/\sqrt{155}$, or 4.5 ± 0.13

For knowledge, we have $35.0 \pm 1.65(6.2)/\sqrt{155}$, or 35.0 ± 0.82

To cover all nurses in community hospitals, we would want our convenience sample to serve as a representative sample, and the four hospitals to accurately represent all community hospitals. For all nurses, we would want the sample to reflect a random sample of all nurses.

7.19 As we are estimating a percentage, we have $\hat{p} = 170/283 = .601$, so the interval is
$.601 \pm 1.56\sqrt{(.601)(.399)/283}$ or $60.1\% \pm 4.5\%$

7.21 $\hat{p} = 86/100 = .86$. A 95% confidence interval is $.86 \pm 1.96\sqrt{.86(.14)/100}$ or $.86 \pm .07$.
Therefore, the smallest fraction is $.86 - .07 = .79$.

7.23 $\hat{p} = 643/1000 = .643$. A 95% confidence interval is $.643 \pm 1.28\sqrt{.643(.357)/100}$ or $.643 \pm .019$.

7.25 Our margin of error is $z\sqrt{\hat{p}\hat{q}/n} = z\sqrt{.37(.63)/1017} = .03$. Solving for z yields $z = 1.98$, and with a central area of .4761, gives 2(.4761) or 95.22% confidence.

7.27 For the major banks, $\Sigma X = 30.21$, $\Sigma X^2 = 106.8149$, $\overline{X} = 3.021$, $s^2 = 1.728$, $s = 1.314$
For the superregional banks, $\Sigma X = 54.92$, $\Sigma X^2 = 309.436$, $\overline{X} = 5.492$, $s^2 = 0.868$, $s = 0.932$

(a) $5.492 \pm 2.26(0.932)/\sqrt{10}$ or 5.492 ± 0.667
(b) $3.021 - 5.492 \pm 2.10\sqrt{s_p^2(1/10 + 1/10)}$ where $s_p^2 = \frac{9(1.728)+9(0.868)}{18} = 1.30$
This reduces to -2.471 ± 1.070

7.29 (a) $120.03 \pm 1.65(31.23)/\sqrt{200}$ or 120.03 ± 3.64
(b) $92.65 \pm 1.65(25.74)/\sqrt{150}$ or 92.65 ± 3.47
(c) $120.03 - 92.65 \pm 1.65\sqrt{\frac{(31.23)^2}{200} + \frac{(25.74)^2}{150}}$ or 27.38 ± 5.03

7.31 $\overline{X}_1 - \overline{X}_2 \pm t\sqrt{s_p^2(\frac{1}{n_1} + \frac{1}{n_2})}$, or $3823 - 5965 \pm 2.06\sqrt{s_p^2(\frac{1}{15} + \frac{1}{12})}$, where
$s_p^2 = \frac{14(1523)^2 + 11(1958)^2}{25}$. This reduces to -2142 ± 1379

7.33 The differences are $-6.6, .1, .9, -3.8, -4.3, 2.3, 2.8, -5.6,$ and -6.8, so $\overline{d} = -2.975$ and $s_d = 3.574$. Therefore, the 95% confidence interval is $-2.975 \pm 2.36(3.574)/\sqrt{8}$ or -2.975 ± 2.988

7.35 First, we need to pair the observations and take the differences. This yields
$-.10, .21, .19, -.21, .02, -.09, .20, -.10, -.01,$ and $-.10$, resulting in $\Sigma D = .01$ and $\Sigma D^2 = .2029$. Therefore, $\overline{D} = .01/10 = .001$ and
$s_D^2 = \frac{.2029 - (.01)^2/10}{9} = .0225$, and $s = .150$.
The interval is $.001 \pm 3.25(.150)/\sqrt{10}$, or $.001 \pm .154$

7.37 As the same people are in each sample, the samples cannot be thought of as independent. We could make a confidence interval, but its validity would be questioned with the problem of independence.

7.39 $\hat{p}_1 = 75/125 = .6$, $\hat{p}_2 = 120/150 = .8$.
The interval is $6 - .8 \pm 1.65\sqrt{\frac{.6(.4)}{125} + \frac{.8(.2)}{150}}$ or $-.2 \pm .09$

7.41 e=w/2=7. n=$(z\sigma/e)^2 = [1.96(20)/7]^2 = 31.36 \rightarrow 32$.

7.43 e=w/2=130/2=65. n=$(z\sigma/e)^2 = [1.96(98)/65]^2 = 8.73 \rightarrow 9$

7.45 This is only a single group problem. n = $z^2pq/e^2 = (2.33)^2(.5)(.5)/.03 = 1508.03 \rightarrow$ 1509.

7.47 $e = z\sqrt{\frac{\sigma^2}{n} + \frac{\sigma^2}{2n}}$, so we have $.5 = 2.58\sqrt{\frac{(2.25)^2}{n} + \frac{(2.25)^2}{2n}} = 2.58\sqrt{7.59/n}$.

Therefore, solving for $\sqrt{n}$ yields $\sqrt{n} = 2.58\sqrt{7.59}/.5 = 14.22$, so n = $(14.22)^2 = 202.21 \rightarrow$ 203.

7.49 n = z^2pq/e^2. With w = .2, e = .2/2 = .1, sp n = $2.33^2(.5)(.5)/(.10)^2 = 135.72 \rightarrow 136$

7.51 w=2, so e=1. n = $z^2(2\sigma^2)/e^2 = 1.88^2(2)(10^2)/1^2 = 706.88 \rightarrow 707$.

7.53 n = $(1.96)^2(.85)(.15)/(.01)^2 = 4898.04 \rightarrow 4899$

7.55 For concessions, we have $\sigma \approx R/4 = 20/4 = 5$. Therefore, n = $z^2\sigma^2/e^2 = (1.96)^25^2/(.5)^2 = 384.16 \rightarrow 385$.

For the five game fractions, n = $z^2pq/e^2 = 1.96^2(.2)(.8)/(.03)^2 = 682.95 \rightarrow 683$. Since we need to satisfy both conditions, we have n = 683.

7.57 Since we don't know s from the study, it is impossible to determine n. The intervals will not overlap, as the first interval goes from 38.6 – 42.6 and the second interval goes from 46.8 to 50.8. Overlapping intervals would indicate a possibility that the true means could be equal, or very close. For them to overlap, we need to make them wider, which requires making n smaller in the denominator.

7.59 For the $\overline{X}$ terms, you should have .165, .230, .265, .198, .218, .293, .593, .230, .373, .230. The R terms should read .13, .26, .44, .18, .18, .38, .25, .18, .63, .18. Therefore, we have $\overline{\overline{X}} = \Sigma\overline{X}/10 = .2795$ and $\overline{R} = .281$.

(a) For the $\overline{X}$-chart, $\overline{\overline{X}} \pm A_2\overline{R}$ yields $.2795 \pm (0.729)(.281)$ or $.2795 \pm .2048$. Therefore, the LCL = .0747 and the UCL = .4843.

For the R-chart, $D_3\overline{R} = 2.282(.281) = .641$, and $D_4\overline{R} = 0$.

(b)

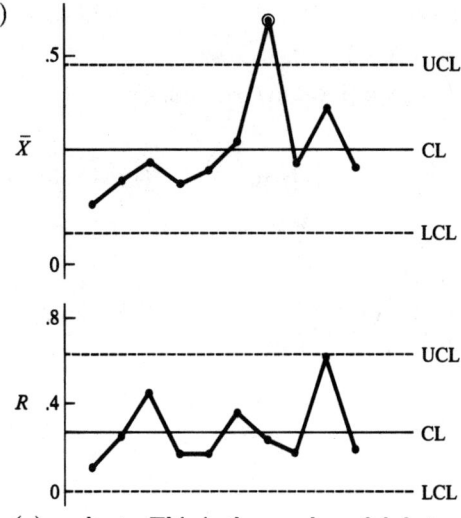

7.61 (a) c chart. This is the number of defects per observation.

(b) $\bar{c} = 29/8 = 3.625$. For a c-chart, we have $3.625 \pm 3\sqrt{3.625}$, or from 0 to 9.337

(c)

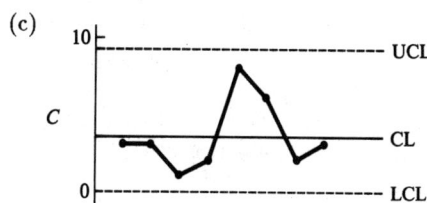

7.63 If we are accurate to within $200, then in repeated sampling, 95% of the estimated average costs will be within $\pm$ $200 of the true average cost. For the problem, we are 95% confident that the mean cost for public sector employees is 2071 ± 200 (or form $1871 - $2271) and for corporate employees, we are 95% confident that the true average cost is between $2164 - $2564. If the individual estimates are within $\pm$$200, then we have $1.96 s/\sqrt{n} = 200$, or $s/\sqrt{n} = 200/1.96 = 102.04$. In estimating $\mu_1 - \mu_2$, the estimation error is

$$1.96\sqrt{\frac{s_1^2}{n_1} + \frac{s_2^2}{n_2}} = 1.96\sqrt{2(102.04)^2} = 282.84$$

7.65 $\hat{p}_1 = 39/120 = .325$, $\hat{p}_2 = 12/80 = .15$

(a) $.325 - .15 \pm 2.05\sqrt{(.325)(.675)/120 + (.15)(.85)/80}$, or $.175 \pm .120$

(b) $.325 \pm 2.05\sqrt{(.325)(.675)/120}$ or $.325 \pm .088$

(c) $.15 \pm 2.05\sqrt{(.15)(.85)/80}$ or $.15 \pm .082$

7.67 $\hat{p} = 65/250 = .26$, $.26 \pm 2.33\sqrt{.26(.74)/250}$, or $.26 \pm .065$

7.69 $\Sigma X = 1282$, $\Sigma X^2 = 83546$, $\overline{X} = 64.1$, $s^2 = 72.09$, $s = 8.49$

The interval is $64.1 \pm t(8.49)/\sqrt{20}$. For 95% confidence, based on 19 degrees of freedom, t=2.09. For 99% confidence, t=2.86. Therefore, the intervals are $64.1 \pm 3.97, 5.43$

7.71 $\Sigma fX = 4445$, $\Sigma fX^2 = 292622.5$, $\overline{X} = 4445/90 = 49.39$, based on 90 df.
$$s^2 = \frac{292622.5 - (4445)^2/90}{89} = 821.22, \; s = 28.66, \text{ based on 89 df.}$$

7.73 $\Sigma X = 5213$, $\Sigma X^2 = 4976289$

(a) $\overline{X} = 5213/6 = 868.83$
(b) $s^2 = \dfrac{4976289 - (5213)^2/6}{5} = 89412.17$
(c) $s = \sqrt{89412.17} = 299.02$
(d) s^2 and s are based on $6 - 1 = 5$ df.
(e) Assuming population normality, we have $868.83 \pm 2.57(299.02)/\sqrt{6}$ or 868.83 ± 313.85

7.75 (a) $\Sigma X = 9697$, $\Sigma X^2 = 11787275$, $\overline{X} = 9697/8 = 1212.13$, $s^2 = \dfrac{11787275 - (9697)^2/8}{7} = 4756.98$, $s = 68.97$

(b) $\overline{X}$ is based on 8 degrees of freedom, s^2 and s are based on $8 - 1 = 7$ degrees of freedom.

(c) $1212.13 \pm 1.90(68.97/\sqrt{8})$, or 1212.13 ± 46.21

7.77 $n = z^2(\sigma_1^2 + \sigma_2^2)/e^2 = 1.96^2(68.97^2 + 42^2)/5^2 = 1002.02 \to 1003$

7.79 A point estimate is a single value. An interval estimate is a range of values.

7.81 $\hat{p} = 65/150 = .433$. The interval is $.433 \pm 2.58\sqrt{(.433)(.567)/150}$ or $.433 \pm .104$

7.83 (a) See Definitions 7.3, 7.4, and 7.5. (b) $\overline{X}$ and $\hat{p}$ are unbiased, efficient, and consistent; s^2 is unbiased and consistent; s is consistent.

7.85 The interval is $33.0 \pm z(2.4)/\sqrt{75}$. For 95%, 99%, and 90% confidence, the z values are 1.96, 2.58, and 1.65. Therefore, the intervals reduce to $33.0 \pm .54, .71, .46$.

7.87 $n = (1.96)^2(200)^2/(1.25)^2 = 9.83 \to 10$.

7.89 (a) $\Sigma fX = 3449$, $\Sigma fX^2 = 288437$, $\overline{X} = 71.85$, $s^2 = 864.08$, $s = 29.40$

(b) $\overline{X}$ is based on 48 df, s is based on 47 df.

(c) $71.85 \pm 1.96(29.40)/\sqrt{48}$, or 71.85 ± 8.32

7.91 $n = (1.65)^2[(.5)(.5)+(.5)(.5)]/(.05)^2 = 544.5 \to 545$ for each group, or a total of 1090.

7.93 $n = (2.17)^2(.5)(.5)/(.06)^2 = 327.01 \to 328$

7.95 $n = (2.58)^2(2.5)^2/(0.5)^2 = 166.41 \to 167$

7.97 $\hat{p} = 183/200 = .915$, $.915 \pm 1.96\sqrt{(.915)(.085)/200}$, or $.915 \pm .039$

7.99 The $\overline{X}$ terms are 1.003, 1.003, 1.005, .995, .993, 1.020, .995, 1.008. The R terms are .04, .03, .01, .03, .04, .15, .03, .03. Therefore, we have
$\overline{\overline{X}} = 1.0028$ and $\overline{R} = .045$
For the $\overline{X}$ chart, we have $1.0028 \pm 0.729(.045)$, for a LCL=.9700 and UCL=1.0356.
For the R-chart, UCL= $2.282(.045) = .103$, and LCL = 0.

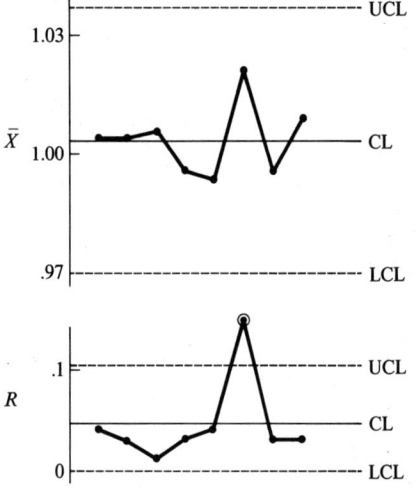

7.101 $\overline{c} = 27/10 = 2.7$. Therefore, we have $2.7 \pm 3\sqrt{2.7}$, or 2.7 ± 4.9. Therefore, UCL=7.6 and LCL=0.

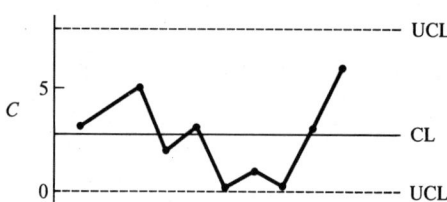

CHAPTER 8

HYPOTHESIS TESTING

8.1 (a) H_0: no leak H_1: leak

(b) A type-one error occurs when we take action to shut down the plant when there is no leak. A type-two error occurs when we do nothing when there is a leak. This could be a major disaster.

(c) Since the type-two error is a potential major disaster, we want a small value for β.

8.3 (a) This is a type-one error b,(c) These are correct decisions (d) This is a type-two error.

8.5 $\alpha = P(X<60 \text{ or } X>70|p=.08) = P(z<\dfrac{59.5-64}{\sqrt{800(.08)(.92)}} \text{ or } z>\dfrac{70.5-64}{\sqrt{800(.08)(.92)}}) =$

$P(z<-0.59 \text{ or } z>0.85) = (.5-.2224) + (.5-.3023) = .4753$

8.7 $\beta = P(X \geq 50|p=.45) = P(z \geq \dfrac{49.5-45}{\sqrt{100(.45)(.55)}}) = P(z \geq 0.90) = .5 - .3159 = .1841$

8.9 Under H_0: p=.3 true, $\mu = 900(.3) = 270$, so we reject H_0 if X<250. Therefore,

$\beta = P(X \geq 250|p=.25) = P(z \geq \dfrac{249.5-225}{\sqrt{900(.25)(.75)}}) = P(z \geq 1.89) = .5 - .4706 = .0294$

8.11 $\beta = P(73<\overline{X}<77|\mu=70) = P(\dfrac{73-70}{25/\sqrt{350}}<z<\dfrac{77-70}{25/\sqrt{350}}) = P(2.24<z<5.24) = .5 - .4875 = .0125$

8.13 H_1: p<.39, $\hat{p} = 60/200 = .3$, $z = \dfrac{.3 - .39}{\sqrt{(.39)(.61)/200}} = -2.61 < -1.65$.

Reject H_0 at $\alpha = .05$. Yes, there has been an improvement.

8.15 At $\alpha = .01$, we want z>2.33. Therefore, z=$\dfrac{\hat{p} - .6}{\sqrt{(.6)(.4)/200}}$>2.33.

Solving for $\hat{p}$ gives $\hat{p}$>.681, or X>.681(200) = 136.2, so $X \geq 137$.

8.17 H_1: p<.53, $z = \dfrac{.48 - .53}{\sqrt{.53(.47)/500}} = -2.24 < -1.64$. Reject H_0 at $\alpha = .05$.

The change has reduced the viewer share.

The p-value is $P(z<-2.24) = .5 - .4875 = .0125$

- 49 -

8.19 (a) H_1: $p>.25$, $\hat{p}=55/200=.275$ $z = \dfrac{.275 - .25}{\sqrt{(.25)(.75)/200}} = 0.82 < 1.65$.

Do not reject H_0 at $\alpha = .05$. There is insufficient evidence of campaign effectiveness.

(b) $p = P(z>0.82) = .5 - .2939 = .2061$

8.21 H_1: $p>.6$. Therefore, $\dfrac{\hat{p} - .6}{\sqrt{(.6)(.4)/900}} > 1.48$, so $\hat{p} = .6242$, or 62.42%

8.23 (a) H_1: $p<1/3$, $\hat{p} = 1939/6265 = .309$

$z = \dfrac{.309 - .333}{\sqrt{.333(.667)/6265}} = -4.00 < z^* = -1.65$ Reject H_0 at $\alpha = .05$.

Wake Forest is most competitive.

(b) No. We could need to conclude that $p \geq 1/3$. With $z = -4$ from part a, this cannot happen.

(c) Yes. If we fail to reject H_0 in (a), we cannot conclude that WF is most competitive, but be cannot conclude that it is not.

(d) With $z = -4.00$, we would reject H_0 for any α. We might set α at .001.

8.25 (a) the standard deviations

(b) $\dfrac{64{,}356 - 60{,}000}{s/\sqrt{100}} > 1.64$, or $s<26560.98$

(c) $\dfrac{61{,}029 - 60{,}000}{s/\sqrt{100}} > 1.64$, or $s<6274.39$

8.27 (a) H_1: $\mu<5.5$, $\sigma = 9$ inches $= 9/12 = .75$ ft. $z = \dfrac{5 - 5.5}{.75/\sqrt{81}} = -6 < -2.33$

Reject H_0 at $\alpha = .01$. The mean height is less than 5.5 feet.

(b) $p = P(z < -6) = 0$

8.29 (a) H_1: $\mu \neq .75$, $\dfrac{\overline{X} - .75}{.08/\sqrt{49}} > 1.64$ or < -1.64. Therefore, $\overline{X} > .769$ or $\overline{X} < .731$

(b) No. .74 is not $< .731$ (from part a)

(c) $z = \dfrac{\overline{X} - .75}{.08/\sqrt{100}} > 1.64$ or < -1.64. Therefore, $\overline{X} > .7706$ or $\overline{X} < .7294$

(d) It increases. We do not require a value of $\overline{X}$ as far from the hypothesized value.

8.31 H_1: $\mu<250$, $t = \dfrac{230 - 250}{35/\sqrt{19}} = -2.49 < t^* = -2.10$

Reject H_0 at $\alpha = .025$. Yes, there has been a significant decrease from the previous average of 250 calories.

8.33 H_1: $\mu>15$, $\Sigma X=203.6$, $\Sigma X^2 = 3490.78$, $\overline{X} = 16.97$, $s = 1.82$

$t = \dfrac{16.97 - 15}{1.82/\sqrt{12}} = 3.75 > t^*=1.80$. Reject H_0 at $\alpha=.05$. The ad is misleading.

8.35 (a) $H_1: \mu_I > \mu_{II}$, $t = \dfrac{125 - 107}{\sqrt{733(1/15 + 1/29)}} = 2.09 > t^* = 1.68$

where $s_p^2 = \dfrac{14(400) + 28(900)}{42} = 733.33$

Reject H_0 at $\alpha = .05$. The union is correct. Population I has higher average benefits than does Population II.

Since $t > t_{.025, 42}$ and $t < t_{.01, 42}$, then $.025 > p > .01$.

8.37 For each variable, we have $H_1: \mu_1 \neq \mu_2$, with $t^* = \pm 2.58$

For ROE, we have $t = \dfrac{11.93 - 14.01}{\sqrt{s_p^2(1/67 + 1/120)}}$, where $s_p^2 = \dfrac{66(9.45)^2 + 119(10.15)^2}{185}$

$t = -1.377$. There is not a significant difference.

For income, we have $t = \dfrac{104.91 - 98.37}{\sqrt{s_p^2(1/67 + 1/121)}}$, where

$s_p^2 = \dfrac{66(182.55)^2 + 120(178.85)^2}{186}$

$t = 0.238$. There is not a significant difference.

For Sales, we have $t = \dfrac{2277.37 - 1677.26}{\sqrt{s_p^2(1/67 + 1/122)}}$, where

$s_p^2 = \dfrac{66(2927.66)^2 + 121(2515.17)^2}{187}$

$t = 1.479$. There is not a significant difference.

For Assets, we have $t = \dfrac{1800.48 - 1493.93}{\sqrt{s_p^2(1/67 + 1/122)}}$, where

$s_p^2 = \dfrac{66(2234.45)^2 + 121(2429.55)^2}{187}$

$t = 0.853$. There is not a significant difference.

For Advertising, we have $t = \dfrac{1.35 - 1.17}{\sqrt{s_p^2(1/67 + 1/1170)}}$, where

$s_p^2 = \dfrac{66(2.48)^2 + 116(2.27)^2}{182}$

$t = 0.500$. There is not a significant difference.

8.39 (a) We have $H_1: \mu_D \neq 0$. The differences are 13.9, 25.5, 13.1, 16.1, 2.7, -0.2, 29.6, and 35.1, yielding $\Sigma D = 135.8$, $\Sigma D^2 = 3389.78$, $\overline{D} = 16.975$, and $s_D = 12.447$.

$t = \dfrac{16.975}{12.447/\sqrt{8}} = 3.86 > t^* = 2.36$. Reject H_0 at $\alpha = .05$.

There is a difference in the mean density between the US and Canada.

(b) $16.975 \pm 2.36(12.447)/\sqrt{8}$, or 16.975 ± 10.41

(c) Normality of the differences, random selection of the pairs

8.41 The differences are 100, −18, 30, 50, −15, −5, 70, 51, 22, 5, 35, 15, 5, 60, 50. With $\Sigma D = 455$ and $\Sigma D^2 = 29559$, $\overline{D} = 30.33$, $s_D = 33.55$.

$H_1: \mu_D \neq 0$, $t = \dfrac{30.33 - 0}{33.55/\sqrt{15}} = 3.50 > t^* = 1.76$

Reject H_0 at $\alpha = .10$. The North has a higher mean profit.

8.43 Subtracting right to left, the differences are 2, 6, 13, −5, 6, 7, 8, and 31. Therefore, $\Sigma D = 68$ and $\Sigma D^2 = 1344$, and $\overline{D} = 8.5$ and $s_D = 10.46$.

$H_1: \mu_D > 0$ (If you subtract left to right, your sign should be <)

$t = \dfrac{8.5 - 0}{10.46/\sqrt{8}} = 2.30 > t^* = 1.90$.

Reject H_0 at $\alpha = .05$. The course is effective in increasing the mean speeds.

8.45 The differences are 7, 7, 11, 7, 1, 12, 13, −2, 2, so $\Sigma D = 58$, $\Sigma D^2 = 590$, $\overline{D} = 6.44$, and $s_D = 5.20$.

$H_1: \mu_D < 10$, $t = \dfrac{6.44 - 10}{5.20/\sqrt{9}} = -2.05 < t^* = -1.40$. Reject H_0 at $\alpha = .10$.

There is sufficient evidence to conclude false claims, i.e., that the average weight loss is less than 10 lb.

8.47 $\hat{p}_1 = 492/1553 = .317$, and $\hat{p}_2 = 312/1279 = .244$. $H_1: p_1 - p_2 > 0$

$z = \dfrac{.317 - .245}{\sqrt{\hat{p}\hat{q}(1/1553 + 1/1279)}}$, where $\hat{p} = \dfrac{492 + 312}{1553 + 1279} = .284$ and $\hat{q} = .716$

$= 4.28 > z^* = 1.65$. Reject H_0 at $\alpha = .05$. There is a significant increase in the proportion of plans with violations. The p-value is $P(z > 4.28) \approx 0$.

8.49 (a) For Dec. 88 vs. Dec. 89, we have $H_1: p_1 - p_2 \neq 0$, with

$z = \dfrac{.337 - .275}{\sqrt{.306(.694)(1/5000 + 1/5000)}} = 6.73 > z^* = 1.96$. Reject H_0 at $\alpha = .05$.

There was a higher percentage in December, 1988.

(Note: the pooled percentage works out to be the average of the two estimates since the sample sizes are equal)

For Dec. 89 vs. Nov. 89, we have $H_1: p_1 - p_2 \neq 0$, with

$z = \dfrac{.275 - .296}{\sqrt{.2855(.7145)(1/5000 + 1/5000)}} = -2.32 < z^* = -1.96$. Reject H_0 at $\alpha = .05$.

There was a higher percentage in November, 1989

(b) For home purchases, For Dec. 88 vs. Dec. 89, we have H_1: $p_1 - p_2 \neq 0$, with
$$z = \frac{.038 - .032}{\sqrt{.035(.965)(1/5000 + 1/5000)}} = 1.63 < z^* = 1.96.$$ Do not reject H_0 at $\alpha = .05$. There was not a significant difference in percentages.

(Note: the pooled percentage works out to be the average of the two estimates since the sample sizes are equal)

For Dec. 89 vs. Nov. 89, we have H_1: $p_1 - p_2 \neq 0$, with
$$z = \frac{.032 - .030}{\sqrt{.031(.969)(1/5000 + 1/5000)}} = 0.58 < z^* = 1.96.$$ Do not reject H_0 at $\alpha = .05$. There was not a significant difference in percentages.

For income, For Dec. 88 vs. Dec. 89, we have H_1: $p_1 - p_2 \neq 0$, with
$$z = \frac{.278 - .295}{\sqrt{.2865(.7135)(1/5000 + 1/5000)}} = -1.88 > z^* = -1.96.$$ Do not reject H_0 at $\alpha = .05$. There was not a significant difference in percentages.

(Note: the pooled percentage works out to be the average of the two estimates since the sample sizes are equal)

For Dec. 89 vs. Nov. 89, we have H_1: $p_1 - p_2 \neq 0$, with
$$z = \frac{.295 - .293}{\sqrt{.294(.706)(1/5000 + 1/5000)}} = 0.22 < z^* = 1.96.$$ Do not reject H_0 at $\alpha = .05$. There was not a significant difference in percentages.

(c) Yes. Month to month, there could be minor differences which are not statistically significant, such as the last example. However, by the time they build on one another, the final difference between first and last month could be significant. For example, if each month increased by .3 to .5%, at the end of the year, the difference might be $3-6\%$ which could be significant.

(d) Yes, as long as the first and last months are not that different. For example, if the first 5 months all went way up, but the next 6 months all went way down, the net change at the end of the year may not be statistically significant.

8.51 $\hat{p}_1 = 19/58 = .328$, $\hat{p}_2 = 22/49 = .449$, $\hat{p}_p = (19+22)/(58+49) = .3832$
H_1: $p_1 \neq p_2$, $z = \dfrac{.328 - .449}{\sqrt{.3832(1/58 + 1/49)}} = -1.28 > z^* = -1.96$

Do not reject H_0 at $\alpha = .05$. There is no evidence of a difference of a difference in the proportions of out-of-court settlements between the two judges.

8.53 $H_1: p_1 > p_2$, $\hat{p}_1 = 29/40 = .725$, $\hat{p} = 20/50 = .4$, $\hat{p}_p = (29+20)/(40+50) = .544$

$z = \dfrac{.725 - .4}{\sqrt{.544(1/40 + 1/50)}} = 3.07 > z^* = 2.33$. Reject H_0 at $\alpha = .01$.

Yes, the percentage of the sufferers claiming relief from the new product is higher than the percentage claiming relief from the placebo.

$p = P(z > 3.07) = .5 - .4989 = .0011$.

8.55 $H_1: p > .5$, $\hat{p} = 154/282 = .546$, $z = \dfrac{.546 - .5}{\sqrt{.5(.5)/282}} = 1.54 > z^* = 1.28$.

Reject H_0 with 90% confidence. The owners have an edge (they win more than half of the cases).

8.57 Null: a,c,f,g,j Alternative: b,d,e,h,i

8.59 No, it doesn't make sense. We could conclude that the medical staff quality was more important, but we can't interpret what the numbers mean.

8.61 $H_1: p_1 < p_2$, $\hat{p}_1 = 245/400 = .613$, $\hat{p}_2 = 550/850 = .647$, $\hat{p}_p = (245+550)/(400+850) = .636$

$z = \dfrac{.613 - .647}{\sqrt{.636(1/400 + 1/850)}} = -1.17 > z^* = -1.64$. Do not reject H_0 at $\alpha = .05$.

No, this is not evidence that the IBM owners are more likely to have dot-matrix printers than Apple owners.

8.63 $\beta = P(\overline{X} < 58 | \mu = 58) = .5$

8.65 $\alpha = P(z > 2 \text{ or } z < -2) = 2(.5 - .4772) = .0456$

8.67 $H_1: p_1 \neq p_2$, $\hat{p}_1 = 63/80 = .788$, $\hat{p}_2 = 39/60 = .65$, $\hat{p}_p = (63+39)/(80+60) = .729$

$z = \dfrac{.788 - .65}{\sqrt{.729(1/80 + 1/60)}} = 1.82 > 1.75$.

Reject H_0 at $\alpha = .08$. Type A bushes have a better chance of survival.

8.69 Subtracting right to left, the differences are

.8,.1,3.4,1.1,.4,.9,.5,1.8,-2.0,2.1,-1.3,1.3,3.2,3.1,1.7,.9,-.3, and 5. This gives $\Sigma D = 22.7$, $\Sigma D^2 = 78.31$, $\overline{D} = 1.26$, and $s_D = 1.71$.

$H_1: \mu_2 > \mu_1$, $t = \dfrac{1.26 - 0}{1.71/\sqrt{18}} = 3.13 > t^* = 1.74$. Reject H_0 at $\alpha = .05$.

The cars have a higher mean gas mileage with the shield.

8.71 $H_1: \mu < 10$, $t = \dfrac{9.5 - 10}{2/\sqrt{25}} = -1.25 > -1.71$. Do not reject H_0 at $\alpha = .05$.

There is not enough evidence to conclude the company was lying.

8.73 (a) mean(breaking strength) (b) one-tail(increase in strength)

(c) $z = \dfrac{55,250 - 55,000}{500/\sqrt{50}} = 3.54$ $P(z > 3.54) = 0$

(d) Yes. $p = 0 < \alpha = .01$. We reject H_0 at $\alpha = .01$.

8.75 $\beta = P(225 \leq \overline{X} \leq 275|\mu=200) = P(\frac{225-200}{75/\sqrt{144}} \leq z \leq \frac{275-200}{75/\sqrt{144}}) = P(4 \leq z \leq 12) = 0$
If $\mu_1 = 275$, $\beta = P(\frac{225-275}{75/\sqrt{144}} \leq z \leq 0) = P(-6 \leq z \leq 0) = .5$

8.77 H_1: $\mu_1 > \mu_2$, $t = \frac{23-18}{\sqrt{23.69(1/9 + 1/6)}} = 1.95 > t^* = 1.77$,

where $s_p^2 = \frac{8(16)+5(36)}{13} = 23.69$

Reject H_0 at $\alpha = .05$. "On the average," men weigh more pounds over their ideal weights than do women.

8.79 (a) The differences are $.7, -.5, .8, -.6, 7.5, 2.3, 0, 3, 3.5$, yielding $\Sigma D = 16.7$, $\Sigma D^2 = 84.53$, $\overline{D} = 1.86$, and $s_D = 2.59$.
H_1: $\mu_D > 0$, $t = \frac{1.86 - 0}{2.59/\sqrt{9}} = 2.15 < t^* = 2.90$ Do not reject H_0 at $\alpha = .01$.
No, there is not evidence that the Type A cap has a longer average opening time over the Type B cap.

(b) For H_1: $\mu < 7$, we have $\overline{X} = 6.53$, $s = 3.02$, and

$t = \frac{6.53 - 7}{3.02/\sqrt{9}} = -0.46 > t^* = -1.90$. Do not reject H_0 at $\alpha = .05$.

No, there is not evidence that the mean time is significantly less than 7 seconds.

8.81 H_1: $p \neq .1$, $\hat{p} = 40/350 = .114$ $z = \frac{.114 - .1}{\sqrt{(.1)(.9)/350}} = 0.89 < z^* = 1.96$
Do not reject H_0 at $\sigma = .05$. At this time, there is no need to change the fraction of left-handed sets.

8.83 H_1: $\mu_1 \neq \mu_2$, $t = \frac{23.5 - 19}{\sqrt{12.5(1/11 + 1/11)}} = 2.98 > t^* = 2.09$

where $s_p^2 = \frac{10(16) + 10(9)}{20}$. Reject H_0 at $\alpha = .05$.

Yes, the groups differ in their opinions. The major corporation CEO's required a higher mean number of vacation days.

8.85 (a) H_0: The check is bad H_1: The check is good

(b) A type-one error exists if we accept a bad check. We risk giving away a free piece of merchandise.

(c) A type-two error exists when we refuse a good check. Not only do we lose a potential sale, but probably lose a future customer as well.

(d) It depends. From a long term point of view, a type-two error is worse. A type-one error may involve only a one-time problem, and the owner might still be able to take legal action to recoup the loss. A type-two error has longer implications.

8.87 The differences, right to left, are 23,2,8,20,5, −4,4,1,5,10,10,15, −11,8, −2,9,8, yielding $\overline{D}$ = 6.53, and s_D = 8.35.
H_1: $\mu_D > 0$ (<0 if you subtract left to right), $t = \frac{6.53 - 0}{8.35/\sqrt{17}} = 3.22 > t^* = 1.75$
Reject H_0 at $\alpha = .05$. The people have higher mean scores after taking the course.

8.89 If we do not reject H_0 for two-tailed test, then the hypothesized value of the parameter should be enclosed in the confidence interval. However, all confidence intervals that we have studied to this point are "two-tailed," whereas many tests are one-tailed. Therefore, some adjustments may be necessary.

CHAPTER 9

THE F DISTRIBUTION AND ANALYSIS OF VARIANCE

9.1 (a) $F_{10,20}$ is a distribution of all possible ratios s_1^2/s_2^2, where s_1^2 and s_2^2 are both unbiased estimates of σ^2, the variance of a normal distribution. s_1^2 is obtained from a sample of size 11 and s_2^2 is obtained from a sample of size 21.

(b) $\mu = 20/18 = 1.11$, the median is less than one since $n_1 < n_2$

9.3 yes: a,e,h no: b,c,d,i can't tell: f,g

9.5 H_1: $\sigma_I^2 > \sigma_{II}^2$, $F = (8.50/6.75)^2 = 1.59 > 1.44 = F_{50,150}$
(We were given standard deviations, so we had to square them for the test. The degrees of freedom are 50 and 178. This is the closest value from the F table)
Reject H_0 at $\alpha = .05$. Area I has higher variation than Area II.

9.7 H_1: $\sigma_A^2 < \sigma_B^2$, $F = (6.9/4.3)^2 = 2.57 > 2.53 = F_{12,14}$
(We were given standard deviations, so we had to square them for the test. The degrees of freedom are 14 and 14. This is the closest value from the F table)
Reject H_0 with 95% confidence. Locker B has more variation than Locker A.

9.9 $s_G^2 = [(1500)^2/20 + (1400)^2/20 + \ldots + (1560)^2/20 - (1500+1400+\ldots+1560)^2/120]/5 = 3797.34$

$s_W^2 = \dfrac{19(100)+19(144)+\ldots+19(121)}{120-6} = 109$

$F = 3797.34/109 = 34.84 > 3.20 = F_{5,100} > F_{5,114}$

Reject H_0 at $\alpha = .01$. There is a difference in the effectiveness of the six processes.

9.11 $\Sigma \overline{X} = 25.28$, $\Sigma \overline{X}^2 = 127.99$ $s_G^2 = 10[\dfrac{127.99 - (25.28)^2/5}{4}] = .4303$

$s_W^2 = \dfrac{9(.05)+9(.01)+9(.07)+9(.06)+9(.10)}{45} = .058$

$F = .4303/.058 = 7.42 > 2.58 = F_{4,44} > F_{4,45}$

Reject H_0 at $\alpha = .05$. There is a difference among the five machines.

9.13 $\Sigma T^2/n = (610^2+574^2+\ldots+590^2)/20 = 94084.60$, $\Sigma T = 3058$

$s_G^2 = \dfrac{94084.6 - (3058)^2/100}{4} = 142.74$, $s_W^2 = \dfrac{19(4.5)^2+\ldots+19(4.1)^2}{95} = 23.56$

$F = 142.74/23.56 = 6.06 > 2.48 = F_{4,80} > F_{4,95}$

Reject H_0 at $\alpha = .05$. There is difference among the methods.

9.15 The model totals are 121.4, 83.4, and 122.2, for a grand total of 327.

$\Sigma T^2/n = (121.4)^2/7 + (83.4)^2/4 + (122.2)^2/6 = 6333.12$

Therefore, $SS_G = 6460.47 - (327)^2/17 = 43.18$

$SS_W = (15.3)^2 + (16.7)^2 + \ldots + (17.0)^2 - 6333.12 = 592.72$

$$F = \frac{43.18/2}{592.72/14} = 0.51 < F_{2,14} = 3.74$$

Do not reject H_0 at $\alpha = .05$. The is no significant differences in the mean down times among the three models.

9.17 (a) Group totals are 25, 21, 30, 36, and 9 for an overall total of 121.

$\Sigma T^2/n = 25^2/7 + 21^2/9 + 30^2/7 + 36^2/8 + 9^2/6 = 442.36$

$SS_G = 442.36 - (121)^2/37 = 46.66$, based on 4 df

$SS_W = 3^2 + 7^2 + 6^2 + \ldots + 3^2 + 4^2 - 442.36 = 106.64$, based on 32 df

$$F = \frac{46.66/4}{106.64/32} = 3.50 > F_{4,32} = 2.67$$

Reject H_0 at $\alpha = .05$. There is a difference among makes in the mean number of service calls.

(b) $SST = 3^2 + 7^2 + 6^2 + \ldots + 3^2 + 4^2 - (121)^2/37 = 153.30 = 46.66 + 106.64$

(c) $36 = 4 + 32$

9.19 (a) We have s^2 for the five groups to be 173.79, 12.54, 31.12, and 165.13.

Therefore, using Equation (9.4), $s_W^2 = \frac{9(173.79) + 9(12.54) + 9(31.12) + 9(165.13)}{36} = 95.80$

Using (9.7), we have $\Sigma \overline{X} = 93.7 + 83.9 + 77.3 + 64.8 = 319.7$, and $\Sigma \overline{X}^2 = 93.7^2 + 83.9^2 + 77.3^2 + 64.8^2 = 25,993.23$

Therefore, $s_G^2 = 10[\frac{25,993.23 - (319.7)^2/4}{3}] = 1470.69$

$F = 1470.69/95.80 = 15.35 > F_{3,36} = 4.38$. Reject H_0 at $\alpha = .01$.

There is a difference among the means of the four groups.

(b) We have $\Sigma X = 3197$, $\Sigma X^2 = 94^2 + 92^2 + \ldots + 60^2 + 51^2 = 263,381$

$\Sigma T^2/n = (937)^2/10 + (839)^2/10 + (773)^2/10 + (648)^2/10 = 259,932.30$

Therefore, $s_G^2 = \frac{259,932.30 - (3197)^2/40}{3} = 1470.69$

$s_W^2 = \frac{263,381 - 259,932.30}{40 - 4} = 95.80$, and $F = 15.35$

9.21 For satisfaction, we have H_0: $\mu_1=\mu_2=\mu_3=\mu_4$, with $\Sigma\overline{X} = 19.29$, $\Sigma\overline{X}^2 = 93.0301$

$$s_G^2 = 422[\frac{93.0301 - (19.29)^2/4}{3}] = .6576$$

$$s_W^2 = \frac{96(1.43)^2 + 33(1.15)^2 + 85(1.17)^2 + 205(1.20)^2}{418} = 1.557$$

$F = .6576/1.557 = 0.422 < F^*_{3,418,.05} = 2.60$

Do not reject H_0 at $\alpha = .05$. There are not significant differences among the groups for mean satisfaction scores.

For absenteeism(short), we have H_0: $\mu_1=\mu_2=\mu_3=\mu_4$, with $\Sigma\overline{X} = 7.52$, $\Sigma\overline{X}^2 = 14.5236$

$$s_G^2 = 422[\frac{14.5236 - (7.52)^2/4}{3}] = 54.297$$

$$s_W^2 = \frac{96(2.09)^2 + 33(2.21)^2 + 85(1.69)^2 + 205(1.91)^2}{418} = 3.757$$

$F = 54.297/3.757 = 14.45 > F^*_{3,418,.05} = 2.60$

Reject H_0 at $\alpha = .05$. There are significant differences among the groups for mean short term absenteeism.

For absenteeism(long), we have H_0: $\mu_1=\mu_2=\mu_3=\mu_4$, with $\Sigma\overline{X} = 4.28$, $\Sigma\overline{X}^2 = 5.0488$

$$s_G^2 = 422[\frac{5.0488 - (4.28)^2/4}{3}] = 66.001$$

$$s_W^2 = \frac{96(5.19)^2 + 33(4.41)^2 + 85(3.14)^2 + 205(3.47)^2}{418} = 15.596$$

$F = 66.001/15.596 = 4.23 > F^*_{3,418,.05} = 2.60$

Reject H_0 at $\alpha = .05$. There are significant differences among the groups for long-term absenteeism means.

For eating habits, we have H_0: $\mu_1=\mu_2=\mu_3=\mu_4$, with $\Sigma\overline{X} = 10.23$, $\Sigma\overline{X}^2 = 26.2207$

$$s_G^2 = 422[\frac{26.2207 - (10.23)^2/4}{3}] = 8.085$$

$$s_W^2 = \frac{96(0.80)^2 + 33(0.58)^2 + 85(0.74)^2 + 205(0.74)^2}{418} = 0.553$$

$F = 8.085/0.553 = 14.61 > F^*_{3,418,.05} = 2.60$

Reject H_0 at $\alpha = .05$. There are significant differences among the groups for mean eating habit scores.

9.23 The group totals are 185, 192, 138, and 151, for a grand total of 666, yielding group averages of 37, 38.4, 27.6, and 30.2. We have

$\Sigma T^2/n = (185^2 + 192^2 + 138^2 + 151^2)/5 = 22{,}586.8$

$SS_G = 22586.8 - (666)^2/20 = 409$

$SS_W = 37^2 + 37^2 + \ldots + 29^2 + 37^2 - 22586.8 = 307.2$

$F = \dfrac{409/3}{307.2/16} = 7.10 > F_{3,16} = 3.24$

Reject H_0 at $\alpha = .05$. There is a difference in the mean number of participants for the four seminars.

To find the differences, for the LSD test, we have $d^* = 2.12\sqrt{19.2(1/5 + 1/5)} = 6.88$. Therefore, any pair of means needs to differ by more than 6.88 in order to be significant. Therefore, we have have Bonds higher than Real Estate, Tax Shelter higher than Real Estate, and Bonds higher than Estate Planning.

9.25 (a) The group totals are 752, 691, and 466 for a grand total of 1909.

$\Sigma X^2 = 149{,}331$, $(\Sigma X)^2/25 = 145{,}771.24$

$\Sigma T^2/n = (752)^2/9 + (691)^2/9 + (466)^2/7 = 146{,}909.51$

$SS_G = 146{,}909.51 - 145{,}771.24 = 1{,}138.27$ based on 2 df.

$SS_W = 149{,}331 - 145{,}771.24 = 2{,}421.49$ based on 22 df.

$F = \dfrac{1138.27/2}{2421.49/22} = 5.17 > F_{2,22} = 3.44$

Reject H_0 at $\alpha = .05$. There is a difference in the mean scores among the commercials.

(b) To use the LSD approach, we have $d^* = 2.07\sqrt{(2421.49/22)(1/9 + 1/n)}$, where n = 7 or 9, depending on whether Commercial III is in the comparison. If n=7, then $d^* = 10.97$. For n=9, $d^* = 10.26$. Therefore, we have

		I	II	III
		83.56	76.78	66.57
I	83.56	0	6.78	16.99**
II	76.78		0	10.21

Commercial I gets higher mean ratings than Commercial III. No other significant differences exist at this time.

(c) Using the ith root approach will only change the t-value from (b) With 3 groups, we have the third root of .95 = .983, so we use .01 to approximate $\alpha/2$. With 22 df, $t_{.01} = 2.51$, so $d^* = 13.27$ for n=7 and 12.41 for n=9. It does not change any conclusions from part b.

9.27 2.57, 2.23, 2.13, 2.09, 2.06; 6.61, 4.96, 4.54, 4.35, 4.24

9.29 For the data, the institution(A) totals are 4324, 2893, and 2480. The funding group(B) totals are 7007, 965, 292, 970, and 473 for an overall total of 9697. We have

$$\frac{(\Sigma X)^2}{n_T} = \frac{(9697)^2}{15} = 6{,}268{,}787.27$$

$$\Sigma \frac{T_A^2}{n_A} = \frac{4324^2}{5} + \frac{2893^2}{5} + \frac{2480^2}{5} = 6{,}643{,}365$$

$$\Sigma \frac{T_B^2}{n_B} = \frac{7007^2}{3} + \frac{965^2}{3} + \frac{292^2}{3} + \frac{970^2}{3} + \frac{463^2}{3} = 17{,}089{,}935.67$$

$$\Sigma X^2 = 2609^2 + 493^2 + \ldots + 2^2 + 18^2 = 17{,}866{,}075$$

$$SS_A = \Sigma \frac{T_A^2}{n_A} - \frac{(\Sigma X)^2}{n_T} = 6{,}643{,}365 - 6{,}268{,}787.27 = 374{,}577.53, \text{ based on 2 df.}$$

$$SS_B = \Sigma \frac{T_B^2}{n_B} - \frac{(\Sigma X)^2}{n_T} = 17{,}089{,}935.67 - 6{,}268{,}787.27 = 10{,}821{,}148.4, \text{ based on 4 df.}$$

$$SS_T = \Sigma X^2 - \frac{(\Sigma X)^2}{n_T} = 17{,}866{,}075 - 6{,}268{,}787.27 = 11{,}597{,}287.73, \text{ based on 14 df.}$$

$$SS_W = SS_T - SS_A - SS_B = 11{,}597{,}287.73 - 374{,}577.53 - 10{,}821{,}148.4 = 40156.6,$$

based on 8 df.

The ANOVA Table is given below:

Source	df	SS	MS	F
Inst(A)	2	374,577.53	187,288.76	3.73
Fund(B)	4	10,821,148.4	2,705,287.1	53.90
Within	8	401,561.6	50,195.2	
Total	14	11,597,287.73		

To test Institutions, we have H_0: $\mu_1 = \mu_2 = \mu_3$, and $F = 3.73 < F^*_{2,8,.05} = 4.46$. Do not reject H_0 at $\alpha = .05$. There is insufficient evidence to show a difference in mean funding among the institutions.

9.31 (a) The lab(A) totals are 67.1, 53.4, 148.1, 58.1, and 76.0. The task(B) totals are 33.8, 16.5, 80.9, 9.1, 28.3, 17.5, 20.5, 42.8, 75.5, and 77.8 for an overall total of 402.7. We have

$$\frac{(\Sigma X)^2}{n_T} = \frac{(402.7)^2}{50} = 3243.35$$

$$\Sigma \frac{T_A^2}{n_A} = \frac{67.1^2}{10} + \frac{53.4^2}{10} + \frac{148.1^2}{10} + \frac{58.1^2}{10} + \frac{76.0^2}{10} = 3843.92$$

$$\Sigma \frac{T_B^2}{n_B} = \frac{33.8^2}{5} + \frac{16.5^2}{5} + \ldots + \frac{77.8^2}{5} = 4630.93$$

$$\Sigma X^2 = 5.8^2 + 2.7^2 + \ldots + 16.8^2 + 15.5^2 = 5737.47$$

$$SS_A = \Sigma \frac{T_A^2}{n_A} - \frac{(\Sigma X)^2}{n_T} = 3843.92 - 3243.35 = 600.57, \text{ based on 4 df.}$$

$$SS_B = \Sigma \frac{T_B^2}{n_B} - \frac{(\Sigma X)^2}{n_T} = 4630.93 - 3243.35 = 1387.58, \text{ based on 9 df.}$$

$$SS_T = \Sigma X^2 - \frac{(\Sigma X)^2}{n_T} = 5737.47 - 3243.35 = 2494.12, \text{ based on 49 df.}$$

$$SS_W = SS_T - SS_A - SS_B = 2492.12 - 600.57 - 1387.58 = 505.97 \text{ based on 36 df.}$$

The ANOVA Table follows:

Source	df	SS	MS	F
Lab(A)	4	600.57	150.14	10.69
Task(B)	9	1387.58	154.18	10.97
Within	36	505.97	14.05	
Total	49	2494.12		

To test Labs, we have H_0: $\mu_1 = \mu_2 = \mu_3 = \mu_4 = \mu_5$, and $F = 10.69 > F^*_{4,36,.05} \approx 2.65$ Reject H_0 at $\alpha = .05$. There is a difference in mean times among the labs.

(b) To find differences among the Labs, using the ith root approach, at $\alpha = .05$, with 5 groups, we have $\binom{5}{2} = 10$ possible comparisons. Therefore, we need a confidence level of $.95^{1/10} = .9949$. Therefore, we have
$$d^* = t_{.0025,36}\sqrt{s_w^2(1/10 + 1/10)} = 2.81\sqrt{14.05(.2)} = 4.71$$

	Lab Types				
	Govt	Sup	All	Univ	Ind
	14.81	7.60	6.71	5.81	5.34
Govt 14.81		7.21*	8.1*	9.0*	9.47*
Sup 7.60			0.89	1.79	2.26
All 6.71				0.90	1.37
Univ 5.81					0.47

Therefore, The government labs had higher mean completion times than the other labs. No other significant differences exist at $\alpha=.05$.

(c) We need to assume population normality for the distribution of times for the labs, equal population variances, and a random selection of tasks.

9.33 We have Oil totals as 189.90, 180, 172.6, 172 for a grand total of 714.5. for the Agents, the totals are 246.3, 238.3, 229.9.

For the oil, we have $\Sigma T^2/n = 21{,}306.13$. For the Agents, we have $\Sigma T^2/n = 21{,}288.07$. Therefore,

$SS_A = 21{,}306.13 - (714.5)^2/24 = 34.87$, based on 2 df.

$SS_B = 21{,}288.07 - (714.5)^2/24 = 16.81$ based on 3 df.

$SS_{AB} = (66.7)^2/2 + (65.1)^2/2 + \ldots + (58.2)^2/2 - (714.5)^2/24 - 34.87 - 16.81 = 21.36$ based on 6 df.

$SS_W = 34.7^2 + 32.0^2 + \ldots + 28.8^2 + 29.4^2 - [(66.7)^2/2 + (65.1)^2/2 + \ldots + (58.2)^2/2]$
$= 14.20$, based on 12 df.

(a) To test for oils, $F = \dfrac{34.87/3}{14.20/12} = 9.82 > F_{3,12} = 3.49$

Reject H_0 at $\alpha = .05$. There is a difference among means when different oils are used.

(b) To test for cleaning agents, $F = \dfrac{16.81/2}{14.20/12} = 7.10 > F_{2,12} = 3.88$

Reject H_0 at $\alpha = .05$. There is a difference among means when different cleaning agents are used.

(c) To test for interaction, $F = \dfrac{21.36/6}{14.20/12} = 3.01 > F_{6,12} = 3.00$

Reject H_0 at $\alpha = .05$. There is an interaction effect between oils and agents

9.35 For the networks, the totals are 59.1, 51.2, and 46.5, for an overall total of 156.8. For the time slots, the totals are 12.9, 11.8, 20.8, 13.6, 10.9, 23.5, 16.4, 22.5, 15.9, and 8.5. We have

$\Sigma X^2 = 951.72$, $(\Sigma X)^2/30 = 819.54$

For the networks, $\Sigma T^2/n = [59.1^2+51.2^2+46.5^2]/10 = 827.65$

For the Time Slots, $\Sigma T^2/n = [12.9^2+11.8^2+...+15.9^2+8.5^2]/3 = 898.19$

$SS_T = 951.72 - 819.54 = 132.18$

$SS(\text{Network}) = 827.65 - 819.54 = 8.11$ based on 2 df.

$SS(\text{Time}) = 898.19 - 819.54 = 78.65$ based on 9 df.

$SS_W = 132.18 - 8.11 - 78.65 = 45.42$, based on 18 df.

To test the networks, $F = \dfrac{8.11/2}{45.42/18} = 1.61 < F_{2,18} = 3.55$

Do not reject H_0 at $\alpha = .05$. There are no significant differences in the mean ratings among the three networks. Therefore, there is no need to do multiple comparisons.

9.37 $\Sigma X^2 = 9^2 + 83^2 + ... 112^2 + 23^2 = 224{,}965$, $\Sigma X = 2679$, $(\Sigma X)^2/n = (2679)^2/48 = 149{,}521.69$

For Shifts, $\Sigma T^2/n = [503^2+923^2+1253^2]/16 = 167{,}184.19$

For pay, $\Sigma T^2/n = [1757^2 + 922^2]/24 = 164{,}047.21$

For interaction, $\Sigma(AB)^2/n = [331^2+172^2+640^2+283^2+786^2+467^2]/8 = 183089.88$

$SS_A = 167{,}184.19 - 149{,}521.69 = 17{,}662.50$ based on 2 df

$SS_B = 164{,}047.21 - 149{,}521.69 = 14{,}525.52$ based on 1 df

$SS_{AB} = 183{,}089.88 - 149{,}521.69 - SS_A - SS_B = 1380.17$ based on 2 df.

$SS_W = 224{,}965 - 183{,}089.88 = 41{,}875.12$ based on 42 df.

For the test on Shifts, $F = \dfrac{17{,}662.5/2}{41{,}875.12/42} = 8.86 > F_{2,42} = 5.15$

For the test on Pay, $F = \dfrac{14{,}525.52/1}{41{,}875.12/42} = 14.57 > F_{1,42} = 7.27$

For the test on Interaction, $F = \dfrac{1{,}380.16/2}{41{,}875.12/42} = 0.69 < F_{2,42} = 5.15$

We conclude that at $\alpha=.05$, there is a difference among means of shifts, there is a difference between means of pay levels, but there is no significant interaction.

9.39 The week totals are 176, 170, and 198, for a grand total of 544. The item totals are 166, 87, 135, 59, 65, and 32. We have

$\Sigma X^2 = 21{,}024$, $(\Sigma X)^2/18 = 16440.89$

For Weeks, $\Sigma T^2/n = [176^2+170^2+198^2]/6 = 16{,}513.33$

For Items, $\Sigma T^2/n = [166^2+87^2+135^2+59^2+65^2+32^2]/3 = 20{,}693.33$

$SS_T = 21{,}024 - 16440.89 = 4{,}583.11$

$SS(\text{Week}) = 16{,}513.33 - 16{,}440.89 = 72.44$, based on 2 df.

$SS(\text{Item}) = 20{,}693.33 - 16{,}440.89 = 4{,}252.44$, based on 5 df.

$SS_W = 4{,}583.11 - 72.44 - 4{,}252.44 = 258.83$, based on 10 df.

To test the weeks, we have $F = \dfrac{72.44/2}{258.83/10} = 1.40 < F_{2,10} = 4.10$

Do not reject H_0 at $\alpha = .05$. There is no significant difference in mean weekly sales at this time.

9.41 (a) For states, the totals are 66924, 121936, and 334725 for a total of 523,585. For years, the totals are 204544, 167413, and 151628.

$SS_T = 22644^2 + 21310^2 + \ldots + 105425^2 + 80500^2 - (523585)^2/9 = 15839972000$.

$SS(\text{States}) = (66924)^2/3 + (121936)^2/3 + (334725)^2/3 - (523585)^2/9 = 13335872800$, based on 2 df.

$SS(\text{Year}) = (204544)^2/3 + (167413)^2/3 + (151628)^2/3 - (523585)^2/9 = 491997824$, based on 2 df.

$SS_W = 15839972000 - 13335872800 - 491997824 = 2012101376$.

This makes for the following ANOVA Table:

Source	df	SS	MS	F
State	2	13335872800	6667936400	13.26
Year	2	491997824	245998912	0.49
Within	4	2012101376	503025344	
Total	8	15839972000		

(b) $F = 13.26 > F_{2,4,.05} = 6.94$

Reject H_0 at $\alpha=.05$. There is a difference in mean net income among states.

(c) $F = 0.49 < F_{2,4,.05} = 6.94$

Do not reject H_0 at $\alpha=.05$. There is not a significant difference in mean net income among years.

9.43 For networks, the totals are 19.5, 19.6, and 25.2, for a grand total of 64.3. For viewer age, the totals are 11.3, 30.4, and 22.6.

$\Sigma X^2 = 535.03$, $(\Sigma X)^2/9 = 459.39$

For Networks, $\Sigma T^2/n = [19.5^2 + 19.6^2 + 25.2^2]/3 = 466.48$

For Age, $\Sigma T^2/n = [11.3^2 + 30.4^2 + 22.6^2]/3 = 520.87$

$SS_T = 535.03 - 459.39 = 75.64$

$SS(\text{Network}) = 466.48 - 459.39 = 7.09$ based on 2 df.

$SS(\text{Age}) = 520.87 - 459.39 = 61.48$ based on 2 df.

$SS_W = 75.64 - 7.09 - 61.48 = 7.07$ based on 4 df.

To test the networks, $F = \dfrac{7.09/2}{7.07/4} = 2.00 < F_{2,4,.05} = 6.94$

Do not reject H_0 at $\alpha=.05$. No, there is not evidence of a difference in mean numbers of viewers among the networks.

9.45 H_1: $\sigma_A^2 > \sigma_B^2$ $F = 31/23 = 1.38 < F_{24,24} = 1.98$. Do not reject H_0 at $\alpha = .05$.

There is not enough evidence to conclude more age variability in MBA managers than for non-MBA managers.

9.47 H_0: $\mu_1=\mu_2=\mu_3=\mu_4=\mu_5=\mu_6$

Group totals are 51,65,189,134,75,54 for a grand total of 568.

$\Sigma T^2/n = (51)^2/5 + (65)^2/4 + ... + (54)^2/4 = 9{,}777.09$

$SS_G = 9{,}777.09 - (568)^2/35 = 559.26$, based on 5 df.

$SS_W = 8^2+11^2+...+13^2+13^2 - 9{,}777.09 = 366.91$, based on 29 df.

$F = \dfrac{559.26/5}{366.91/29} = 8.84 > F_{5,29} = 3.73$ Reject H_0 at $\alpha=.01$

There is a difference in the mean outputs among the six age groups.

9.49 Let I, II, and III represent Gloss Glow, Glide Easy, And Sparkle, and C, P, and O represent Cedar, Pine, and Oak. We have

	IC	IP	IO	IIC	IIP	IIO	IIIC	IIIP	IIIO	TOTALS
ΣX	19.3	21.2	17.7	21.0	20.9	23.6	19.3	22.3	20.5	185.8
ΣX^2	98.73		82.35		109.95		94.95		108.11	
		117.66		112.7		141.6		127.35		993.4
n	4	4	4	4	4	4	4	4	4	36
$(\Sigma X^2)/n$		112.36		110.25		139.24		124.3225		
		93.1225	78.3225		109.2025		93.1225		105.0625	965.00
$\overline{X}$	4.825	5.3	4.425	5.25	5.225	5.9	4.825	5.575	5.125	

$SS_G = 965 - (185.8)^2/36 = 6.07056$

For Paints, group totals are 58.2, 65.5, and 62.1. For Surfaces, the totals are 59.6, 64.4, and 61.8.

$SS_A = (58.2^2+65.5^2+62.1^2)/12 - (185.8)^2/36 = 2.22389$ based on 2 df.

$SS_B = (59.6^2+64.4^2+61.8^2)/12 - (185.8)^2/36 = 0.96222$ based on 2 df.

$SS_{AB} = 965 - (185.8)^2/36 - SS_A - SS_B = 2.88445$ based on 4 df.

$SS_W = 4.2^2+5.8^2+...+5.9^2+6.0^2 - 965 = 28.395$, based on 27 df.

For the test on Paint, $F = \dfrac{2.22389/2}{28.395/27} = 1.06 < F_{2,27} = 3.35$

For the test on Surface, $F = \dfrac{0.96222/2}{28.395/27} = 0.46 < F_{2,27} = 3.35$

For the test on Interaction, $F = \dfrac{2.88445/4}{28.395/27} = 0.69 < F_{4,27} = 2.73$

Do not reject H_0 at $\alpha=.05$ for all three tests. No differences or interaction has been shown. The FTC should advise Gloss Glow that their claims have not been substantiated.

9.51 (a) The group totals are 52.71, 90.06, 123.44, 40.35, and 31.25, for a grand total of 337.81.

$\Sigma X^2 = 10.84^2+10.30^2 + ... + 11.80^2+10.95^2 = 3843.3929$

$\Sigma T^2/n = (52.71)^2/5 + (90.06)^2/8 + (123.44)^2/10 + (40.35)^2/4 + (31.25)^2/3 = 3825.81$

$SS_G = 3825.8141 - (337.81)^2/30 = 21.9609$ based on 4 df.

$SS_W = 3843.3929 - 3825.8141 = 17.6788$ based on 25 df.

$F = \dfrac{21.9609/4}{17.6788/25} = 7.76 > F_{4,25} = 2.76$. Reject H_0 at $\alpha = .05$.

There is a difference in mean hourly pays among the 5 contribution categories.

(b) $SS_T = 39.6397 = 21.9609 + 17.6788$

(c) $df_T = 30 - 1 = 29 = 4 + 25$

9.53 The values above which .05(.01) of the terms lie are the values needed for a two tailed test at $\alpha = .10(.02)$.

9.55 For the differences to be significant, they need to be higher than
$$d^* = t_{10}\sqrt{3.59(1/6 + 1/6)}$$
To use the ith root approach, at $\alpha = .05$, we want $\alpha = 1 - (.95)^{1/3} = .017$, so we will use .01 for $\alpha/2$. Therefore,
$$d^* = 2.76\sqrt{3.59(1/6 + 1/6)} = 3.02$$

	Company		
	I	II	III
	28.83	24.67	24.83
I	28.83	0	4.16**	4.00**
II	24.67		0	−0.17

Therefore, Company I has higher average salaries than do Companies II and III.

9.57 $H_1: \sigma_2^2 > \sigma_1^2$, $F = 87.56/41.35 = 2.12 < F_{7,21} = 2.49$
Do not reject H_0 at $\alpha=.05$. There is insufficient evidence to conclude that the supplement significantly reduces weight gain variability.

9.59 The station totals are 108, 131, 96, and 83 for a grant total of 418.
$\Sigma T^2/n = (108^2+131^2+96^2+83^2)/8 = 5{,}616.25$
$SS_G = 5{,}616.25 - (418)^2/32 = 156.125$ based on 3 df.
$SS_W = 15^2+17^2+...+11^2+9^2 - 5616.25 = 1997.75$, based on 28 df.
$$F = \frac{156.125/3}{1997.75/28} = 0.73 < F_{3,28} = 2.95$$
Do not reject H_0 at $\alpha=.05$. There are no significant differences in the mean numbers of commercial minutes among the four stations.

9.61 (a) The group totals are 457, 343, 226, and 363, for a grand total of 1389.
$\Sigma T^2/n = (457)^2/6 + (343)^2/5 + (226)^2/4 + (363)^2/5 = 97460.77$
$(\Sigma X)^2/20 = (1389)^2/20 = 96466.03$
$\Sigma X^2 = 84^2+93^2+...+72^2+53^2 = 101121$
$SS_G = 97460.77 - 96466.03 = 994.72$ based on 3 df.
$SS_W = 101121 - 97460.77 = 3660.23$ based on 16 df.
$$F = \frac{994.72/3}{3660.23/16} = 1.45 < F_{3,16} = 3.24$$
Do not reject H_0 at $\alpha = .05$. There is not a significant difference in mean stress quotients among the four managerial levels.

(b) $s_G^2 = 994.72/3 = 331.57$, based on 3 df. $s_W^2 = 3660.23/16 = 228.76$, based on 16 df.
$s_T^2 = 4654.95/19 = 244.997$ based on 19 df.

(c) $SS_T = 4654.95 = 994.72 + 3660.23 = SS_G + SS_W$

$df_T = 19 = 3 + 16 = df_G + df_W$

9.63

	AH	AL	BH	BL	CH	CL	DH	DL	TOTALS
ΣX	1751	2152	1390	2176	1462	2133	1448	2169	14681
ΣX^2	779199	1211790	497716	1221370	560262	1151227	553390	1227119	7202073
n	4	4	4	4	4	4	4	4	32
$(\Sigma X)^2/n$	766500.25		483025		534361		524166		6963144.6
		1157776		1183744		1137422.2		1176140.2	
$\overline{X}$	437.75	538	347.5	544	365.5	533.25	362	542.25	

SS_A(Package) $= [39030^2 + 3566^2 + 3595^2 + 3617^2]/8 - (14681)^2/32 = 9192.3$ based on 3 df.

SS_B(Price) $= [6051^2 + 8530^2]/24 - (14681)^2/32 = 207851.2$ based on 1 df.

$SS_{AB} = 6963144.6 - (14681)^2/32 - SS_A - SS_B = 10733.6$ based on 3 df.

$SS_W = 427^2 + 483^2 + ... + 485^2 + 702^2 - 6963144.6 = 238928.4$ based on 24 df.

(a) To test the packages, $F = \dfrac{9192.3/3}{238928.4/24} = 0.31 < F_{3,24} = 3.01$

Do not reject H_0 at $\alpha = .05$. There is no significant difference in mean sales among the four package types.

(b) To test the prices, $F = \dfrac{207851.2/1}{238928.4/24} = 20.88 > F_{1,24} = 4.26$

Reject H_0 at $\alpha = .05$. Mean sales differ between the two price categories.

(c) To test the interaction, $F = \dfrac{10733.6/3}{238928.4/24} = 0.36 < F_{3,24} = 3.01$

Do not reject H_0 at $\alpha = .05$. There is no significant interaction between package types and price categories.

9.65 $\Sigma \overline{X} = 107.9$, $\Sigma \overline{X}^2 = 2929.81$, $s_G^2 = 20[\dfrac{2929.81 - (107.9)^2/4}{3}] = 128.05$

$s_W^2 = \dfrac{19(46.5) + 19(58) + 19(48) + 19(51)}{76} = 50.875$

$F = 128.05/50.875 = 2.52 < 2.72 = F_{3,80} < F_{3,76,.05}$

Do not reject H_0 at $\alpha = .05$. The mean layoffs per factory do not differ significantly among the four industries.

9.67 (a) The group totals are 35, 25, and 18, for an overall total of 78. We have

$\Sigma X^2 = 6^2 + 7^2 + ... + 3^2 + 2^2 = 446$

$\Sigma T^2/n = [35^2 + 25^2 + 18^2]/5 = 434.8$

$SS_G = 434.8 - (78)^2/15 = 29.2$, based on 2 df.

$SS_W = 446 - 434.8 = 11.2$ based on 12 df.

$F = \dfrac{29.2/2}{11.2/12} = 15.64 > F_{2,12} = 3.88$

Reject H_0 at $\alpha = .05$. There is a difference in the mean numbers of reported accidents per highway.

(b) For the ith root, our confidence level is $(.95)^{1/3} = .9830$, so $\alpha = .017$ and $\alpha/2 = .0085$, and the closest value in our t-tables is .01. Therefore, we have

$$d^* = t_{.01,12}\sqrt{.93(1/5 + 1/5)} = 2.68\sqrt{.93(1/5 + 1/5)} = 1.638$$

	I	II	III
	7	5	3.6
I 7	0	2**	3.4**
II 5		0	1.4

Therefore, Road I has a higher mean number of reported accidents than both II and III. There is no significant difference between Highways II and III.

9.69 (a)

					Totals
ΣX	24.5	15.0	17.0	27.4	83.9
ΣX^2	123.01	50.40	61.24	156.66	391.31
n	5	5	5	5	20
$(\Sigma X)^2/n$	120.05	45	57.8	150.152	373.002
$\bar{X}$	4.9	3.0	3.4	5.48	

$SS_G = 373.002 - (83.9)^2/20 = 21.0415$, $s_G^2 = 21.0415/3 = 7.0138$

$SS_W = 391.31 - 373.002 = 18.308$, $df_W = 16$, $s_W^2 = 1.14425$

$F = 7.0138/1.14425 = 6.13 > F_{3,16}^* = 3.24$. Reject H_0 at $\alpha = .05$. There are differences among the banks.

(b) $d^* = 2.12\sqrt{1.14425(1/5 + 1/5)} = 1.43$

	4.9	3.0	3.4	5.48
4.9	0	1.9**	1.5**	0.58
3.0		0	0.4	2.48**
3.4			0	2.08**
5.48				0

Security is significantly higher than both Bank and Trust and United.

Piggy Bank is significantly higher than both Bank and Trust and United.

No other differences are significant.

9.71 The group totals are 38, 56, 51, and 72 for a grand total of 217.

$\Sigma T^2/n = (38)^2/4 + (56)^2/6 + (51)^2/4 + (72)^2/7 - (217)^2/21 = 2274.49$

$\Sigma X^2 = 8^2 + 12^2 + \ldots + 14^2 + 12^2 = 2373$

$SS_G = 2274.49 - (217)^2/21 = 32.155$ based on 3 df.

$SS_W = 2373 - 2274.49 = 88.51$ based on 17 df.

$H_0: \mu_1 = \mu_2 = \mu_3 = \mu_4$, $F = \dfrac{32.155/3}{88.51/17} = 2.06 < F_{3,17} = 3.20$.

Do not reject H_0 at $\alpha = .05$. There is no evidence of a difference in the mean heights among the four fertilizers.

9.73 City totals are 11.10, 20.70, and 9.20 for a grand total of 41. Item totals are 4.03, 4.46, 12.47, 4.12, 11.79, and 4.13. We have

$\Sigma X^2 = 158.59$, $(\Sigma X)^2/18 = 93.39$

For Cities, $\Sigma T^2/n = [11.1^2 + 20.7^2 + 9.2^2]/6 = 106.06$

For Items, $\Sigma T^2/n = [4.03^2 + \ldots + 4.13^2]/3 = 121.56$

$SS_T = 158.59 - 93.39 = 65.20$

$SS(\text{Cities}) = 106.06 - 93.39 = 12.67$, based on 2 df.

$SS(\text{Item}) = 121.56 - 93.39 = 28.17$, based on 5 df.

$SS_W = 65.20 - 12.67 - 28.17 = 24.36$, based on 10 df.

(a) To test the Cities, $F = \dfrac{12.67/2}{24.36/10} = 2.60 < F_{2,10,.05} = 4.10$

Do not reject H_0 at $\alpha = .05$. There is no evidence of a difference in mean prices among the three cities.

(b) For London, $\overline{X}_2 = 20.7/6 = 3.45$; for Mexico City, $\overline{X}_3 = 9.20/6 = 1.53$.

The interval is $3.45 - 1.53 \pm 2.23\sqrt{(24.36/10)(1/6 + 1/6)}$ or 1.92 ± 2.01. Since 0 is contained in the interval, there is no significant difference in the mean prices between London and Mexico City.

9.75 (a) Soda totals are 2461, 2333, and 937 for a grand total of 5731. Week totals are 1203, 1222, 928, 1393, and 985. We have

$SS_T = 543^2+468^2+...+238^2+136^2 - (5731)^2/15 = 347,880.93$

$SS(Soda) = [2461^1+2333^2+937^2]/5 - (5731)^2/15 = 285,851.73$ based on 2 df.

$SS(Week) = [1203^2+1222^2+928^2+1393^2+985^2]/3 - (5731)^2/15 = 47,826.27$ based on 4 df.

$SS_W = 347,880.93 - 285,851.73 - 47,826.27 = 14202.93$, based on 8 df. The ANOVA is

Source	df	SS	MS	F
Soda	2	285,851.73	142,925.86	80.50
Week	4	47,826.27	11,956.57	6.73
Within	8	14,202.93	1,775.37	
Total	14	347880.93		

$F_{2,8,.05} = 4.46$. There are differences in mean sales among the three soda brands.
$F_{4,8,.05} = 3.84$. There are differences in mean sales among the weeks.

(b) For differences to be significant, $d^* = 2.31\sqrt{1775.37(1/5 + 1/5)} = 61.45$

		Coca-Cola	Pepsi-Cola	Seven-Up
		492.2	466.6	187.4
Coca-Cola	492.2	0	25.6	304.8**
Pepsi-Cola	466.6		0	289.2**

Both Coke and Pepsi have higher mean sales than Seven-Up.

CHAPTER 10

SIMPLE REGRESSION AND CORRELATION

10.1 (a)

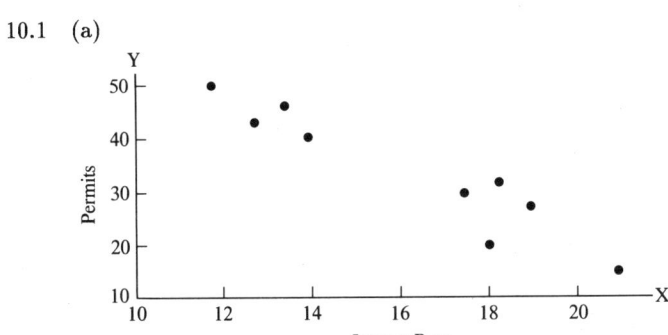

(b) see scattergram

(c) about 33

(d) $\Sigma X=145.75$, $\Sigma Y=302$, $\Sigma X^2=2,444.1875$, $\Sigma Y^2=11,252$, $\Sigma XY=4600.25$; $SS_{xy}= -290.472$, $SS_x=83.847$, and $SS_y=1118.222$.

(e) $b = -290.472/83.857 = -3.464$, $a = 302/9 - (-3.464)(145.75/9) = 89.658$.

Therefore, the best line is $\hat{Y} = 89.658 - 3.464X$

(f) $\hat{Y} = 89.658 - 3.464(16) = 34.234$ or approximately 34.

10.3 (a)

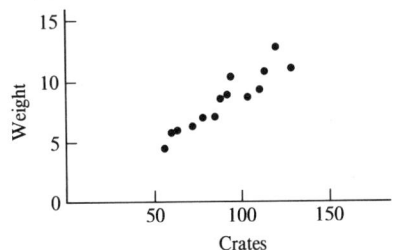

Number of crates is the independent variable.

(b) $\Sigma X=1457$, $\Sigma Y=130.2$, $\Sigma X^2=153,061$, $\Sigma Y^2=1247.46$, $\Sigma XY=13,759.3$; $SS_{xy}=1,112.54$, $SS_x=11,537.733$, and $SS_y=117.324$

(c) b= 1,112.54/11,537.733 = 0.096, a = 130.2/15 − 0.096(1457/15) = −0.645

Therefore, the best line is $\hat{Y} = -0.645 + 0.096X$. The slope is 0.096 and the intercept is −0.645. For each additional crate, we predict that shipment weight increases by 0.096 tons.

(d) $\hat{Y} = -0.645 + 0.096(75) = 6.555$ tons. The estimated mean weight of all 75-crate shipments is 6.555 tons.

10.5 For simplicity, we have divided all data by 1000 to make your calculations simpler.

$\Sigma X = 178.8$, $\Sigma Y = 537.2$, $\Sigma X^2 = 3,396.92$, $\Sigma Y^2 = 29,693.86$, $\Sigma XY = 9,798.89$; $SS_{xy} = 1,794.610$, $SS_x = 732.8$, $SS_y = 5,645.207$, b=2.449, a=8.2766.

The line is $\hat{Y} = 8.2766 + 2.449X$, so with X=15, we have $\hat{Y} = 8.2766 + 2.449(15) = 45.0116$, or 45,002 copies.

10.7 (a) $\Sigma X = 279$, $\Sigma Y = 75.2$, $\Sigma X^2 = 5975$, $\Sigma XY = 1513.8$, $SS_{xy} = 115.080$, $SS_x = 785.6$, b=0.146, and a=2.288. Therefore,

$\hat{Y} = 2.288 + 0.146X$

(b) No. With higher ratings as we get further past the hour, there should be higher rates as the time past hour increases.

(c) At 7 PM, $\hat{Y}=2.28$. At 7:20, $\hat{Y} = 2.288 + 0.146(20) = 5.20$. At 7:25, $\hat{Y} = 2.288 + 0.146(25) = 5.938$.

If costs are proportional to ratings, then at 7:20, the cost should be $(5.20/2.28)(50,000) = \$114,035.09$, and the cost at 7:25 should be $(5.938/2.28)(50,000) = \$130,219.30$

10.9 (a)

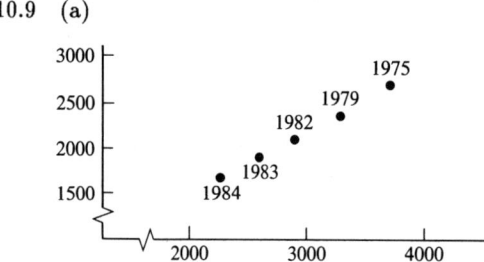

(b) We have $\Sigma X = 14542$, $\Sigma Y = 10624$, $\Sigma X^2 = 43368466$, $\Sigma Y^2 = 23144508$, and $\Sigma XY = 31680373$, $SS_{xy} = 781,531.4$, $SS_x = 1,074,513.2$, and $SS_y = 570,632.8$. b = $781,531.4/1,074,513.2 = .727$, and a = $(10624/5) - .727(14542/5) = 9.418$
Therefore, we have $\hat{Y} = 9.418 + .727X$. To see how well the model does, we substitute the X values for each year in the equation. Therefore, we have

$X = 3,609 \Rightarrow \hat{Y} = 9.418 + .727(3,609) = 2,633.16$

$X = 3,209 \Rightarrow \hat{Y} = 9.418 + .727(3,209) = 2,342.36$

$X = 2,852 \Rightarrow \hat{Y} = 9.418 + .727(2,852) = 2,082.82$

$X = 2,584 \Rightarrow \hat{Y} = 9.418 + .727(2,584) = 1,887.99$

$X = 2,288 \Rightarrow \hat{Y} = 9.418 + .727(2,288) = 1,672.79$

10.11 $\Sigma X = 70.1$, $\Sigma Y = 207.8$, $\Sigma X^2 = 5430.12$, $\Sigma Y^2 = 886.53$, $\Sigma XY = 1902.79$, $SS_{xy} = 578.54$, $SS_x = 1504.59$, $SS_y = 439.80$, b = 0.3485, a = -0.8911.
SSE = $439.80 - 0.3485(578.54) = 217.35$, $s_\epsilon^2 = 217.35/9 = 24.15$, $s_\epsilon = 4.914$

$H_1: \beta > 0$, $t = \dfrac{0.3485}{4.914/\sqrt{1504.59}} = 3.035 > t^*_{.05,9} = 1.83$

Reject H_0 at $\alpha = .05$. Spending has a positive linear effect on market share.

10.13 Recall that for Exercise 10.1, that $SS_{xy} = -290.472$, $SS_x = 83.847$, and $SS_y = 1118.222$.
  (a) SSE = $1118.222 - (-3.46)(-290.472) = 112.027$
  (b) $s_\epsilon^2 = 112.027/7 = 16.004$, $s_\epsilon = 4.00$
  (c) $r^2 = \dfrac{1118.222 - 112.027}{1118.222} = 0.90$
  (d) $r = -\sqrt{0.90} = -0.949$
  (e) $H_1: \beta \neq 0$, $t = \dfrac{-3.46}{4.00/\sqrt{83.847}} = -7.93 < -t^*_{.025,7} = -2.36$. Reject H_0 at $\alpha = .05$.

The number of permit applications decreases linearly as the interest rates are rising.

10.15 Recall from Exercise 10.3 that $SS_{xy} = 1,112.54$, $SS_x = 11,537.733$, and $SS_y = 117.324$.
  (a) SSE = $117.324 - 0.096(1,112.54) = 10.52$
  (b) $s_\epsilon^2 = 10.52/13 = 0.809$, $s_\epsilon = 0.90$
  (c) $r^2 = \dfrac{117.324 - 10.52}{117.324} = 0.910$
  (d) $r = \sqrt{0.910} = 0.954$
  (e) $H_1: \beta > 0$, $t = \dfrac{0.096}{0.90/\sqrt{11537.733}} = 11.457 > t^*_{.05,13} = 1.78$. Reject H_0 at $\alpha = .05$.

Shipment weights increase linearly as the number of crates increase.

10.17 To test $H_1: \beta>0$, we first need to find s_ϵ. We had from Exercise 10.5 $SS_{xy}=1,794.610$, $SS_x=732.8$, $SS_y=5,645.207$, and $b=2.249$. Therefore,

$SSE = 5,645.207 - (2.249)(1,794.610) = 1250.207$, $s_\epsilon^2 = 1250.207/10 = 125.021$, and $s_\epsilon = 11.18$. Now,

$$t = \frac{2.249}{11.18/\sqrt{732.8}} = 5.929 > t^*_{.05,10} = 1.81. \text{ Reject } H_0 \text{ at } \alpha=.05.$$

The number of copies sold increases linearly as the advertising budget increases.

$$r^2 = \frac{5,645.207 - 1250.207}{5,645.207} = .777$$

10.19 $\Sigma X=1400$, $\Sigma Y=24.4$, $\Sigma X^2=282,500$, $\Sigma Y^2=82.28$, and $\Sigma XY=4392.5$. $SS_{xy}=122.5$, $SS_x=37,500$, $SS_y=7.86$, $b=.0033$.

$SSE = 7.86 - .0033(122.5) = 7.46$, $s_\epsilon^2 = 7.46/6 = 1.24$, $s_\epsilon = 1.11$.

(a) $r^2 = \dfrac{7.86 - 7.46}{7.86} = .051$

(b) $h^2 = 1 - r^2 = .949$

(c) $H_1: \beta \neq 0$, $t = \dfrac{.0033}{1.11/\sqrt{37,500}} = 0.58 < t^* = 1.94$. Do not reject H_0 at $\alpha=.10$.

There is insufficient evidence of a linear relationship.

10.21 $SS_{xy}=395.2$, $SS_x=364.6$, $SS_y=1596.4$, $b=1.084$, $SSE=1596.4 - (1.084)(395.2) = 1,167.78$.

$s_\epsilon^2 = 1,167.78/8 = 145.97$, $s_\epsilon = 12.08$.

(a) $H_1: \beta \neq 0$, $t = \dfrac{1.084}{12.08/\sqrt{364.6}} = 1.71 < t^* = 3.36$. Do not reject H_0 at $\alpha=.01$.

(b) $r = \dfrac{395.2}{\sqrt{(364.6)(1596.4)}} = .518$, $r^2 = .518^2 = .268$

10.23 $\Sigma X=460$, $\Sigma Y=1113$, $\Sigma X^2=22154$, $\Sigma Y^2=130261$, $\Sigma XY=53489$, $SS_{xy}=2291$, $SS_x=994$, $SS_y=6384.1$.

$$r = \frac{2291}{\sqrt{(6384.1)(994)}} = .909$$

10.25 The manager would want to set the price at the high level and go for the highest profit. The customer might as well buy the cheapest diskettes.

10.27 $\Sigma X=1222$, $\Sigma Y=438$, $\Sigma X^2=88572$, $\Sigma Y^2=13026$, $\Sigma XY=33254$. $SS_{xy}=3518.67$, $SS_x=5611.78$, $SS_y=2368$, $b=0.627$, $a=-18.233$.

(a) $\hat{Y} = -18.233 + 0.627X$

(b) SSE = 2368 − (0.627)(3518.67) = 161.79, s_ϵ^2 = 161.79/16 = 10.11, s_ϵ = 3.18

$$\hat{Y} = -18.233 + 0.627(60) \pm 2.12(3.18)\sqrt{1 + 1/18 + \frac{(60 - 67.89)^2}{5611.78}}$$
19.387 ± 6.964

$$\hat{Y} = -18.233 + 0.627(75) \pm 2.12(3.18)\sqrt{1 + 1/18 + \frac{(75 - 67.89)^2}{5611.78}}$$
28.792 ± 6.955

$$\hat{Y} = -18.233 + 0.627(90) \pm 2.12(3.18)\sqrt{1 + 1/18 + \frac{(90 - 67.89)^2}{5611.78}}$$
38.197 ± 7.206

10.29 ΣX=70, ΣY=60, ΣX^2=1025, ΣY^2=733.5, ΣXY=858.75; SS_{xy}=18.75, SS_x=45, SS_y=13.5.

(a) b=0.417, a=6.162

$\hat{Y}$=6.162 + 0.417X

(b) $\hat{Y}$=6.162 + 0.417(10.0) = 10.33

$\hat{Y}$=6.162 + 0.417(8.5) = 9.71

$\hat{Y}$=6.162 + 0.417(15.0) = 12.42

$\hat{Y}$=6.162 + 0.417(12.5) = 11.38

(c) 13 − 10.33 = 2.67 days

(d) First we need to find s_ϵ. SSE = 13.5 − (0.417)(18.75) = 5.68, s_ϵ^2 = 5.68/3 = 1.89, and s_ϵ = 1.38. The 90% confidence interval for the mean number of days missed for all laborers with 10 years experience is

$$10.33 \pm 2.35(1.38)\sqrt{1/5 + \frac{(10 - 14)^2}{45}}, \text{ or } 10.33 \pm 2.42$$

10.31 ΣX=424, ΣY=47, ΣX^2=37188, ΣY^2=483, ΣXY=3774. SS_{xy}=−211.6, SS_x=1232.8, SS_y=41.2, b=−0.17, a=23.82.

Therefore, $\hat{Y} = 23.82 - 0.17(X)$

SSE=41.2 − (−0.17)(−211.6)=5.23, s_ϵ^2 = 5.23/3 = 1.74, s_ϵ = 1.32

(a) X=80, $\hat{Y}$= 23.82 − 0.17(80) = 10.22 ± 3.18(1.32)$\sqrt{1/5 + \frac{(80 - 84.8)^2}{1232.8}}$

or 10.22 ± 1.96, or from 8.26 to 12.18

(b) 10.22 ± 3.18(1.32)$\sqrt{1 + 1/5 + \frac{(80 - 84.8)^2}{1232.8}}$

or 10.22 ± 4.64, or from 5.58 to 14.86

10.33 $\Sigma X=1010$, $\Sigma Y=36.04$, $\Sigma X^2=119{,}530$, $\Sigma Y^2=129.9886$, $\Sigma XY=3849.14$, $SS_{xy}=540.01$, $SS_x=26{,}793.64$, $SS_y=11.91$, $b=.0202$, $a=1.42$.

Therefore, $\hat{Y}=1.42 + .0202X$.

$SSE = 11.91 - .0202(540.01) = 1.002$, $s_\epsilon^2 = 1.002/9 = .111$, and $s_\epsilon = .334$

For $X=90$, $\hat{Y} = 3.238$

(a) $3.238 \pm 1.83(.334)\sqrt{1/11 + (90-91.8)^2/26793.64}$

or $3.238 \pm .184$

(b) $3.238 \pm 1.83(.334)\sqrt{1 + 1/11 + (90-91.8)^2/26793.64}$

or $3.238 \pm .638$

10.35 $\Sigma X=57$, $\Sigma Y=395$, $\Sigma X^2=851$, $\Sigma Y^2=851$, $\Sigma XY=1881$, $SS_{xy}=-1335.43$, $SS_x=386.86$, $SS_y=5385.71$.

(a) $b = -1335.43/386.86 = -3.45$, $a = 56.43 - (-3.45)(8.14) = 84.51$, so $\hat{Y} = 84.51 - 3.45X$

(b) First, we need to find s_ϵ. $SSE = 5385.71 - (-3.45)(-1335.43) = 778.48$, $s_\epsilon^2 = 778.48/5 = 155.70$, and $s_\epsilon = 12.48$.

$X=1.5$, $\hat{Y} = 84.51 - 3.45(1.5) = 79.34 \pm 2.57(12.48)\sqrt{1 + 1/7 + \frac{(1.5-8.14)^2}{386.86}}$, or 79.34 ± 35.97, or from 43.37 to 115.31 (note that the upper limit is theoretically 100)

(c) $\hat{Y}=0$ when $X=24.5$. Giving up on some present debts should increase the odds of future debts.

10.37 $\Sigma X = -108.7$, $\Sigma Y = -156.7$, $\Sigma X^2 = 3427.29$, $\Sigma Y^2 = 7070.75$, and $\Sigma XY = 4357.07$. $SS_{xy} = 950.412$, $SS_x = 11064.152$, and $SS_y = 2159.772$. Therefore,

$r = \dfrac{950.412}{\sqrt{(11064.152)(2159.772)}} = .629$

For H_1: $\rho \neq 0$, we have $t = \dfrac{.629\sqrt{5-2}}{\sqrt{1-.629^2}} = 1.39 < t^*_{3,.025} = 3.18$

Do not reject H_0 at $\alpha = .05$. There is insufficient evidence to show a significant correlation exists.

10.39 H_1: $\rho>0$, $r=.55$, $n=83$. $t = \dfrac{.55\sqrt{81}}{\sqrt{1-.55^2}} = 5.93 > t^* = 2.33$. Reject H_0 at $\alpha=.01$.

The mean length stay per patient increases as the nurse/patient ratio increases.

10.41 H_1: $\rho > .52$, $n=100$, $r=.60$

$w = (1/2)\ln[(1+.6)/(1-.6)] = .69314718$

$\mu_w = (1/2)\ln[(1+.52)/(1-.52)] = .57633975$

$\sigma_w = \sqrt{1/97} = .10153462$

$z = \dfrac{.693 - .576}{.1015} = 1.15 < 2.33$ Do not reject H_0 at $\alpha=.01$.

There is not evidence that ρ exceeds .52.

10.43 Due to the presence of $\sqrt{n-2}$ in the numerator of the test, the test is not only a function of r, but of n as well. Therefore, r=.5 will not be significant for n=7 or smaller, but r=.4 will be significant for n=27 or higher. The significance can also be a function of whether the test is one or two-tailed.

10.45 No, but they need some criteria for admission. If we could accurately predict how well students will do (GPA), then we wouldn't need to actually examine the students at all!

10.47 (a) $\Sigma X = 117890$, $\Sigma Y = 565746$, $\Sigma X^2 = 3116136870$, $\Sigma Y^2 = 66296791484$, $\Sigma XY = 14205664217$. $SS_{xy} = 866505029$, $SS_x = 336526450$, $SS_y = 2283084180.8$, $b = 2.575$, and $a = 52439.39$. Therefore, $X = 25{,}000 \Rightarrow \hat{Y} = 52{,}439.39 + 2.575(25{,}000) = 116{,}814.39 \approx 116{,}814$

(b)

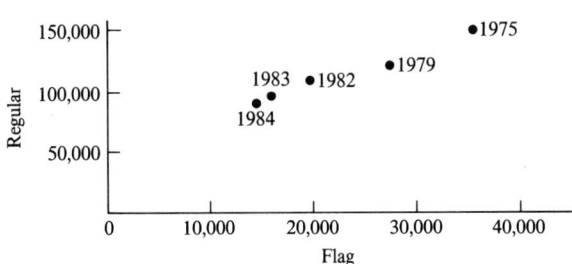

It appears that both the regular and flag services are decreasing over the years.

(c) The flag data appears to be close to a lower limit. Once this occurs, it will not be significantly changing, even if the regular service is changing.

10.49 $\Sigma X = 121$, $\Sigma \log Y = 2.0403231$, $\Sigma X^2 = 1959$, and $\Sigma X \log Y = 52.715953$; $SS_{xy} = 21.856066$, $SS_x = 128.875$, $b = .1695912$, $a = -2.3100265$.

Therefore, $c = 10^{-2.3100265} = .00489749$, and $d = 10^{.1695912} = 1.4777168$, and $\hat{Y} = .0049(1.4777)^X$. When $X = 18$, $\hat{Y} = .0049(1.4777)^{18} = 5.530$, or \$5,530.

10.51 $\Sigma X=13.8$, $\Sigma \log Y=53.672177$, $\Sigma X^2=12.96$, and $\Sigma X \log Y=49.25446$; $SS_{xy}=-0.1239428$, $SS_x=0.264$, $b=-0.4694797$, $a=4.0100664$.
Therefore, $c=10^{4.0100664}=10{,}234.495$, and $d=10^{-0.4694797}=0.33925035$, and $\hat{Y}=10{,}234.495(0.33925)^X$. When $X=1$, $\hat{Y}=3{,}472$

10.53 $\Sigma X=4578$, $\Sigma Y=583$, $\Sigma X^2=2654264$, $\Sigma Y^2=38719$, and $\Sigma XY=302137$. $SS_{xy}=-35239.60$, $SS_x=558455.60$, $SS_y=4730.1$, $b=-.063$, and $a=29.41$.

(a) $\hat{Y}=29.41-.063X$

(b) $s_\epsilon^2 = [4730.1-(-0.063)(-35239.6)]/8 = 313.30$, $s_\epsilon = 17.70$
$H_1: \beta \neq 0$, $t = \dfrac{0.063}{17.70/\sqrt{558455.6}} = 2.26 < t^*_{.025,8} = 2.31$

Do not reject H_0 at $\alpha=.05$. There is not evidence of a linear relationship between the numbers of CD's and tape devices sold.

(c) $X=550$, $\hat{Y}=29.41+.063(550)=64.06$, so the interval is
$64.06 \pm 2.31(17.70)\sqrt{1+1/10+(550-457.8)^2/558455.6}$, or
64.06 ± 43.10

(d) If X is unavailable, then we use $\overline{Y}=58.3$

(e) From before, we have $s_\epsilon = 17.70$

(f) We have $s = \sqrt{SS_y/n-1} = \sqrt{4730.1/9} = 22.93$, somewhat larger than s_ϵ.

10.55 $\Sigma X=108$, $\Sigma Y=28588$, $\Sigma X^2=1200$, $\Sigma Y^2=89092788$, and $\Sigma XY=319803$. $SS_{xy}=11052.60$, $SS_x=33.60$, $SS_y=7{,}365{,}413.6$, $b=328.9464$, and $a=-693.8211$.
$SSE = 7{,}365{,}413.6 - 328.9464(11052.60) = 3{,}729{,}700.62$
$s_\epsilon^2 = SSE/8 = 466{,}212.58$ and $s_\epsilon = 682.80$
$H_1: \beta>0$, $t = \dfrac{328.9464}{682.80/\sqrt{33.60}} = 2.79 > t^* = 1.86$

Reject H_0 at $\alpha=.05$. The number of homes on the market increases linearly with each point rise in the interest rate.

A 95% interval for β is $328.9464 \pm 2.31(682.60)/\sqrt{33.60}$, or 328.9464 ± 271.55

10.57 $\mu_w = (1/2)\ln(1.55/.45) = .61938131$, and $\sigma_w = \sqrt{1/61} = .12803688$
Therefore, $z = \dfrac{w-.61838131}{.12803688} = 1.65$, and $w = .829642 = (1/2)\ln(\dfrac{1+r}{1-r})$.
Therefore, $\dfrac{1-r}{1+r} = e^{1.659284}$, and solving for r gives $r=.68028373$

10.59 (a)

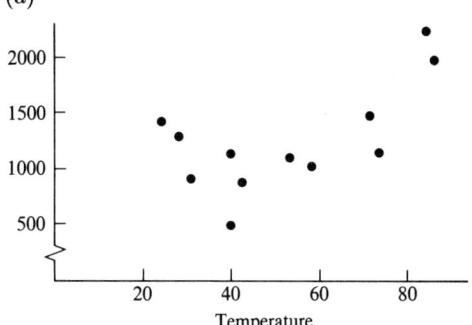

(b) $\Sigma X=630$, $\Sigma Y=15180$, $\Sigma X^2=38158$, $\Sigma Y^2=21,766,180$, and $\Sigma XY=870450$. $SS_{xy}=73,500$, $SS_x=5083$, $SS_y=2,563,480$, $b=14.46$, and $a=505.85$. Therefore, the line is $\hat{Y} = 505.85 + 14.46X$.

(c) SSE = $2,563,480 - 14.46(73,500) = 1,500,670$, so s_ϵ^2 = SSE/10 = 150,067 and s_ϵ = 387.38

$H_1: \beta \neq 0$, $t = \dfrac{14.46}{387.48/\sqrt{5083}} = 2.66 > t^* = 2.23$

Reject H_0 at $\alpha=.05$. Usage is going up as the temperature increases.

(d) For X=50, we have $\hat{Y} = 505.85 + 14.46(50) = 1228.85$.

For X=73, we have $\hat{Y} = 505.85 + 14.46(73) = 1561.43$

For X=29, we have $\hat{Y} = 505.85 + 14.46(29) = 925.19$

(e) Independence may be lacking, as X is a time-related variable.

10.61 (a)

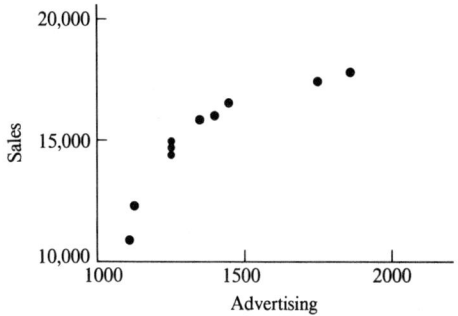

The relationship appears to be exponential.

(b) $\Sigma X=13870$, $\Sigma \log Y=41.8243$, $\Sigma X^2=19743212$, $\Sigma X \log Y=57984.243$, $SS_{xy}=112.639$, $SS_x=505522$, b=.000223, a=3.863383, d=1.0005, and c=7301.008, so that $\hat{Y} = 7301.008(1.0005)^X$. For X=1550, $\hat{Y} = 7301.008(1.0005)^{1550} = 15,844.44$

10.63 No. It may "help," but years of experience is probably a larger influence.

10.65 positive: 1,3,6,9,10; negative: 2,4,8; not correlated: 5,7.

10.67 $\beta=\rho\sigma_x/\sigma_y = .58(4)/5 = .464$

10.69 $\Sigma X = 14.22$, $\Sigma Y = 2641$, $\Sigma X^2 = 21.7132$, $\Sigma Y^2 = 466073$, and $\Sigma XY = 2417.65$, so that $SS_{xy} = -86.018$, $SS_x = 8.233$, $SS_y = 1080.93$. b = -10.45 and a = 185.97. Therefore, for X=2, we predict $\hat{Y} = 185.97 - 10.45(2) = 165.07$(thousands of dollars).

10.71 $\Sigma X=700000$, $\Sigma Y=1498$, $\Sigma X^2=41186000000$, $\Sigma Y^2=235404$ and $\Sigma XY=91100300$. $SS_{xy}=3716966.67$, $SS_x=352666666.70$, and $SS_y=48403.67$

$$r = \frac{3716966.67}{\sqrt{(352666666.70)(48403.67)}} = .8996$$

10.73 perfectly and negatively correlated

10.75 $\overline{Y} = \$215$, and $\overline{X}=1.75$. We have $\hat{Y} = a + bX$, so $295 = a + b(2.5)$, and $a = \overline{Y} - b\overline{X}$, so a = 215 − b(1.75). Therefore, we have 295 = (215 − 1.75b) + 2.5b, so .75b = 80 and b=80.67 and a = 215 − (80.67)(1.75) = 28.33. Therefore, $\hat{Y} = 28.33 + 106.67X$, so X=0.50 gives $\hat{Y} = 81.67$.

10.77 $\Sigma X=713$, $\Sigma Y=61.7$, $\Sigma X^2=47265$, $\Sigma Y^2=390.61$, and $\Sigma XY=3897.3$. $SS_{xy}= -95.50$, $SS_x=1049.64$, $SS_y=44.53$.

$$r = \frac{-95.50}{\sqrt{(1049.64)(44.53)}} = -0.44$$

$H_1: \rho \neq 0$, $t = \frac{-0.44\sqrt{9}}{\sqrt{1-0.44^2}} = -1.47 < t^*=2.26$. Do not reject H_0.

Sam's position is not refuted.

10.79 $\Sigma X=270$, $\Sigma Y=162$, $\Sigma X^2=18600$, $\Sigma Y^2=5888$, $\Sigma XY=10,017$. $SS_{xy}=1269$, $SS_x=4020$, and $SS_y=639.2$.

(a) $r = \dfrac{1269}{\sqrt{4020(639.2)}} = 0.79$

(b) Then, $\Sigma X=320$, $\Sigma X^2=24,500$, and $\Sigma XY=11637$. Then, $SS_{xy}=1269$, and $SS_x=4020$, and r=0.79.

(c) Then $\Sigma Y=187$, $\Sigma Y^2 = 7633$, $\Sigma XY=11,367$, and $SS_{xy}=1269$, $SS_y=639.2$, and r=0.79.

(d) Then $\Sigma X=320$, $\Sigma X^2=24,500$, $\Sigma Y=187$, $\Sigma Y^2=7633$, and $\Sigma XY=13237$. Then, $SS_{xy}=1269$, $SS_x=4020$, and $SS_y=639.2$, and r=0.79

(e) Adding constants to the data does not change the correlation(or the slope of the line).

(f) Multiplying the data by constants will not change the correlation. It will increase SS_{xy} by a factor of 5(10)=50, and SS_x and SS_y by the factor of 10^2=100 and 5^2=25, but these increase will be canceled in the formula for r.

10.81 A prediction interval is an interval for a single value. When we are concerned about individual values, this is a short-term problem. Alternatively, a confidence interval for $\mu_{Y|X}$ is an estimate for an average, which is a long-term parameter.

CHAPTER 11

MULTIPLE REGRESSION AND CORRELATION

11.1 (a) We predict the salary (in $1000) for programmers as $15 + 1.5$ times the number of years of college $+ 2$ times the number of years of experience. With $b_1 = 1.5$, each additional year of college adds an additional 1.5 or $1500 to the predicted salary. With $b_2 = 2$, each additional year of programming experience is worth an additional 2 or $2000 in predicted salary.

(b) $\hat{Y} = -1.37 + 1.5(4) + 2(2) = 25$, or $25,000

11.3 (a) $\hat{Y} = 77.74 + 22.955X_1 - 1.4314X_2$

(b) $b_1 = 22.955$; each additional salesperson accounts for an estimated 22.955 more products on the average per week.

$b_2 = -1.4314$; each additional dollar increase in price gives a predicted reduction of 1.4314 sales.

(c) $\hat{Y} = 77.74 + 22.955(3) - 1.4312(40) = 89.349$ or ≈ 89 sales.

11.5 (a) It increases it by $b_1 = .84$

(b) It increases it by $b_2 = .66$

11.7 $\hat{Y} = 67.1 - 19.1(1.5) - .235(55) = 25.52$

11.9 (a) $s_\epsilon = 10.91$, $R^2 = .897$

(b) Using our formulas from Chapter 10, we have $r^2_{X_1 X_2} = .009$, $r^2_{X_1 Y} = .794$, and $r^2_{X_2 Y} = .056$.

(c) With the small correlation between X_1 and X_2, the assumptions of independence of the X terms appears reasonable. Additionally, it can be concluded that the number of salespeople contributes more to reducing the variation of sales than does the price of the item.

11.11 $s_\epsilon = 3.766$, $R^2 = .80$. The sample standard deviation of the points about the regression line is 3.766, based on 7 degrees of freedom. 80% of the variation of the sample energy consumption can be explained by the linear relationships with the thermostat temperature and the outdoor temperature.

11.13 (a) For H_1: at least one of the β's is not 0, we have

$$F = \frac{.950/2}{.05/(5-1-2)} = 19 > F^*_{2,2,.05} = 9.55$$

Reject H_0 at $\alpha = .05$. At least one of the variables makes a significant contribution to the model.

(b) At this stage, we need to make individual tests for H_0: $\beta_1=0$ and H_0: $\beta_2=0$.

11.15 For H_1: $\xi^2 > 0$, we have $R^2 = .727$.

$$F = \frac{.727/2}{.273/5} = 6.658 > F^*_{2,5,.05} = 5.79$$

Reject H_0 at $\alpha = .05$. The coefficient is significant.

11.17 For H_0: $\beta_1=\beta_2=0$, we have

$$F = \frac{.866/2}{.134/6} = 19.388 > F^*_{2,6,.01} = 10.92. \text{ Reject } H_0 \text{ at } \alpha=.01.$$

At least one of the variables, weight and speed, makes a contribution to the mileage.

11.19 (a)

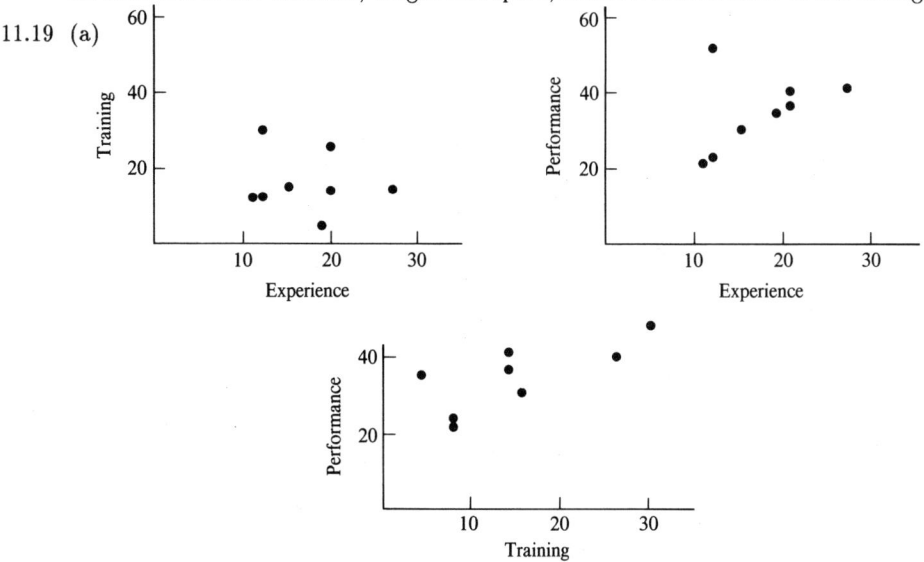

(b) Using the correlation formula from Chapter 10, we have $r^2_{X_1 X_2} = 0$.
(c) Using the correlation formula from Chapter 10, we have $r^2_{X_1 Y} = .134$
(d) Using the correlation formula from Chapter 10, we have $r^2_{X_2 Y} = .593$

11.21 With $r_{X_1 X_2} = -0.8956$, there is potential multicollinearity.

11.23 We need dummy variables for City and Sex. For City, we might have $D_1 = 1$ if New York, 0 if not, $D_2 = 1$ if Boston, 0 if not, and $D_3 = 1$ if Los Angeles, 0 if not. For Sex, we have $D_4 = 1$ if Male, 0 if not.

11.25 For H_0: $\xi^2 = 0$, we have $R^2 = .843$, and

$$F = \frac{.843/3}{.157/18} = 32.22 > F^*_{3,18,.05} = 3.16. \text{ Reject } H_0 \text{ at } \alpha=.05.$$

At least one of the variables contributes to the model. To make the individual tests, we choose to make one-tailed tests, since intuitively, there would appear to be positive relationships, before we ever collected the data.

For H_1: $\beta_1 > 0$, p=.133/2 = .066 > α = .05. Do not reject H_0 at α = .05. The number of offices is not a significant factor.

For H_1: $\beta_2 > 0$, p=.000/2 = .000 < α = .05. Reject H_0 at α = .05. The number of employees has a positive influence on gross sales.

For H_1: $\beta_3 > 0$, p=.051/2 = .025 < α = .05. Reject H_0 at α = .05. The corporate percentage has a positive influence on gross sales.

11.27 (a) Average Price
 (b) $\hat{Y} = 144 + 67.4(\text{Rating}) + 1166.8(\text{TV-homes})$
 (c) Pittsburgh: $144 + 67.4(15) + 1166.8(1.59) = 3010.21$
    Los Angeles: $144 + 67.4(8) + 1166.8(4.6) = 6050.48$

11.29 (a) $k - 1 = 7 - 1 = 6$
 (b) If the F-test in ANOVA is significant, multiple comparisons should reveal exactly which categories differ. If all of the categories are not significant, we can save degrees of freedom by combining the groups.

11.31 It is necessary to have a software package in order to do them. Using Plant Type as a dummy variable(1=T, 0=A), the MINITAB printout yields $\hat{Y} = 280392 + 217987(\text{Plant}) + 611.9(\text{Employees})$, with $R^2 = .515$. With the test for H_0: $\xi^2=0$ yielding a p-value of .009, we know at least one of the variables is significant. However, when we examine these individually, Plant Types has a p-value of .469, which is not significant, while Employees, with p=.007, is significant. Therefore, the best model will not include Plant Type, but just the linear relationship between Employees and Production.

11.33 (a) $s_\epsilon = 3.282$ (b) $R^2 = .45$
 (c) Using our formulas from Chapter 10, we have $r^2_{X_1 X_2} = .0309$, $r^2_{X_1 Y} = .3365$, and $r^2_{X_2 Y} = .1879$.
 (d) With $r^2_{X_1 X_2} = .0309$, multicollinearity does not appear to be a major problem.

11.35 It means that the variable has an estimated inverse (or negative) linear effect on the dependent variable, and we will predict a decrease in y for each unit increase of the variable.

11.37 (a) $H_1: \xi^2 > 0$ $F = \dfrac{.926/2}{.074/9} = 56.31 > F_{2,9,.01} = 8.02$. Reject H_0 at $\alpha = .01$.

The coefficient is significant.

(b) Yes. We need to make individual tests for $H_0: \beta_1 = 0$ and $H_0: \beta_2 = 0$.

11.39 (a) Dispersion about the regression model,

(b) The proportion of total variation explained by the multiple relationship.

11.41 $s_\epsilon = 1.881$, $R^2 = .601$

11.43 As an overall test, it saves a lot of work. If we fail to reject H_0, there is no need to make individual tests.

11.45 $s_\epsilon = 0.2690$

11.47 (a) Using the formula from Chapter 10, we have $r_{X_1 X_2} = -.02$, $r_{X_1 Y} = .81$, and $r_{X_2 Y} = -.56$.

(b) At $\alpha = .10$, we have $t^* = 1.44$. We have

$t = \dfrac{-.02\sqrt{6}}{\sqrt{1 - (.02)^2}} = -0.49$. There is not a significant relationship.

$t = \dfrac{.81\sqrt{6}}{\sqrt{1 - (.81)^2}} = 3.38$. Salt has a positive linear relationship with production rate.

$t = \dfrac{-.56\sqrt{6}}{\sqrt{1 - (.56)^2}} = -1.66$. Chlorine Dioxide has an inverse linear relationship with production rate.

11.49 $\hat{Y} = 31.727 - .13682(\text{invitations}) + .2317(\text{items})$. For $X_1 = 103$ and $X_2 = 6$, $\hat{Y} = 31.727 - .13682(103) + .2317(6) = 19.03 \approx 19$.

11.51 For $H_1: \xi^2 > 0$, and $R^2 = .57$, $F = \dfrac{.57/2}{.43/7} = 4.64 < F_{2,9,.01} = 9.55$

Do not reject H_0 at $\alpha = .01$. There is not evidence of a significant relationship. No further tests are needed.

11.53 $\hat{Y} = -3808 + .387(\text{income}) - 3.1(\text{mortgage}) - 819(\text{dependents})$. For \$50,000 income, 4 dependents, and a mortgage of \$850, we have $\hat{Y} = -3808 + .387(50,000) - 3.1(850) - 819(4) = 9,631$.

11.55 It is possible that only one variable is important. Alternatively, you may have a multicollinearity problem. Tabulate correlations among the independent variables and run a stepwise regression.

11.57 It would indicate if potential relationships should be linear, non-linear, or non-existent. A preliminary graph of independent variables will indicate potential multicollinearity problems.

CHAPTER 12

TIME SERIES AND FORECASTING

12.1 (a) To find the moving averages, we start with $(229+338+330+320)/4 = 304.25$, $(338+330+320+312)/4 = 325$, and so on, producing 354, 373.25, 396, 412.5, 427.25, 451.5, 473.5, 492.75, 516.75, 540.75, 559.25, 568, 594, 610.25, and 622.25. To find the centered moving averages(TC terms), we have $(354+373.25)/2 = 314.625$, $(325+354)/2 = 339.5$, and so on, producing 363.625, 384.625, 404.250, 419.875, 439.375, 462.5, 483.125, 504.75, 528.75, 550, 563.625, 581, 602.125, and 616.250.

(b) We have n=16, $\Sigma TC = 7558$, and $\Sigma X(TC) = 71,104$, and
$$b = \frac{71,104 - (8.5)(7558)}{15(16)(17)/12} = 20.17941,$$ and $a = (7558/16) - 20.179141(8.5) = 300.85$
Therefore, we have $T = 300.85 + 20.17X$. Substituting 1,2,...,16 for X gives 321.03, 341.21, 361.39, 381.57, 401.75, 421.93, 442.11, 462.29, 482.46, 502.64, 522.82, 543, 563.18, 583.36, 603.54, 623.72.

(c) To find the C terms, we divide the TC terms by the T terms, yielding $314.625/321.03 = .980$, $339.5/341.21 = .995$, and so on, producing 1.006, 1.008, 1.006, .995, .994, 1, 1.001, 1.004, 1.011, 1.013, 1.001, .996, .998, and .988.

12.3 To find the moving averages, we start with $(18.3+22.5+26.2+28.8)/4 = 23.95$, $(22.5+26.2+28.8+27.8)/4 = 23.325$, and so on, producing 28.885, 31.375, 34.35, 37.4, 39.85, 42.375, 44.925, 46.925, 49.4, 53.125, 54.6, 57.55, 59.875, 61.25, 64.375, 65.325, 67.775, 71.525, and 73.875. To find the centered moving averages(TC terms), we have $(23.95+26.325)/2 = 25.14$, $(26.325+28.885)/2 = 27.61$. and so on, producing 30.13, 32.86, 35.88, 38.63, 41.11, 43.65, 45.93, 48.16, 51.26, 53.86, 56.08, 58.71, 60.56, 62.81, 64.85, 66.55, 69.65, and 72.70.

We have n=20, $\Sigma TC = 986.13$, and $\Sigma X(TC) = 11997.53$, and
$$b = \frac{11,997.53 - (10.5)(986.13)}{19(20)(21)/12} = 2.4709,$$ and $a = (986.13/20) - 2.4709(10.5) = 23.36$
Therefore, we have $T = 23.36 + 2.4709X$. Substituting 1,2,...,20 for X gives 25.83, 28.30, 30.77, 33.24, 35.71, 38.19, 40.66, 43.13, 45.60, 48.07, 50.54, 53.01, 55.48, 57.95, 60.43, 62.89, 65.37, 67.84, 70.31, and 72.78.

To find the C terms, we divide the TC terms by the T terms, yielding $25.14/25.83 = .973$, $27.61/28.3 = .975$, and so on, producing .979, .989, 1.005, 1.012, 1.011, 1.012, 1.007, 1.002, 1.014, 1.016, 1.011, 1.013, 1.002, .999, .992, .981, .991, and .999.

12.5 (a) For men, the moving averages will be 4.3, 4.3, 4.335, 4.125, 3.975, and 3.75, for TC terms of 4.3, 4.2625, 4.175, 4.05, and 3.8625. With n=5, we have $\Sigma TC = 20.65$, $\Sigma X(TC) = 60.8625$, $SS_{xy} = -1.0875$, $SS_x = 10$, $b = -0.10875$, $a = 4.13 - (-0.10875)(3) = 4.45625$. Therefore, the trend line is $T = 4.45625 - 0.10875X$.

For women, the moving averages will be 5.275, 5.125, 4.875, 4.575, 4.375, and 4.25, for TC terms of 5.2125, 5.0125, 4.725, 4.475, and 4.3125. With n=5, we have $\Sigma TC = 23.7375$, $\Sigma X(TC) = 68.875$, $SS_{xy} = -2.3375$, $SS_x = 10$, $b = -0.23375$, $a = 4.7475 - (-0.23375)(3) = 5.44875$. Therefore, the trend line is $T = 5.44875 - 0.23375X$.

(b) To solve the equation for men, we have $4.45625 - 0.10875X = 0$, so X=41(July, 1996). For women, we have $0 = 5.44875 - 0.23375X$, so X=24(April, 1992).

(c) After the points in part b, predicted unemployment will be negative. If we chart the data for a longer period, we would expect rates to either level off or generate a non-linear trend.

12.9 If we make our moving averages over 4 years, we have 24.25, 25.75, 28.5, ..., 152.75, 151, and 134.75. This yields TC terms of 25, 27.125, 33.25, ..., 150.125, 151.875, and 142.875. We have n=34, $\Sigma TC = 3151$, and $\Sigma X(TC) = 72249.6$, so that $b = 5.228$ and $a = 1.195$. With $T = 1.195 + 5.228X$, we substitute X=1,2,...,33,34 and get 6.423, 11.65, ..., 173.7, and 178.9.

To generate the C terms, we take TC/T = $25/6.423 = 3.892$, $27.125/11.65 = 2.328$, 1.970, ... , .8910, .8743, and .7984.

For the SI terms, we have Y/TC = $21/25 = .84$, $28/27.125 = 1.032$, .7820, ..., 1.306, 1.021, and .826. Since we took a four way average, to find potential seasonal terms, we average out every fourth term, eliminating the high and low terms. Multiplying by the adjustment factor to make everything add to four yields 1.031, .917, 1.028, and 1.024 for the four terms. We now have I = SI/S = $.84/1.031 = .8148$, $1.032/.9168 = 1.126$, .7605, ..., 1.275, .990, and .901.

While we find the seasonal component, there really is not one, as the data is on an annual basis, and there is no apparent seasonal pattern in the data. The large irregular components appear as a function of the wide variation in the data, especially from the large increase between 1964 and 1965. This makes the trend line very inaccurate and causes more variation at later stages of the model.

12.11 To generate the SI term, we take Y/TC, so we have $330/314.625 = 1.049$, $320/339.5 = .943$, and so on, yielding .858, 1.180, 1.007, .979, .860, 1.109, 1.043, .989, .861, 1.107, 1.065, .986, .814, and 1.157. To find the seasonal indices, we average every fourth observation, eliminating the high and low, yielding $(.858+.860)/2 = .859$, $(1.109+1.157)/2 = 1.133$, 1.046, and .983. As these sum to 4.021, rather than 4. we multiply each term by the adjustment factor of $4/4.021 = .995$, so we have $.995(.859) = .855$, $.995(1.133) = 1.127$, 1.041, and .978. Now, we find the irregular components by SI/S, for $1.049/1.041 = 1.008$, $.943/.978 = .964$, and so on, yielding 1.004, 1.047, .967, 1.001, 1.006, .984, 1.002, 1.011, 1.007, .982, 1.023, 1.008, .952, and 1.027.

12.13 To generate the SI term, we take Y/TC, so we have $26.2/25.14 = 1.042$, $28.8/27.61 = 1.043$, and so on, yielding .923, .995, 1.009, 1.054, .973, .974, 1.008, 1.057, .936, .973, 1.091, .967, .987, .982, 1.029, 1.041, .913, and .983. To find the seasonal indices, we average every fourth observation, eliminating the high and low, yielding .944, .990, 1.027, and 1.046. As these total to 3.997, we must multiply each term by the adjustment factor of $4/3.997 = 1.001$, yielding .945, .981, 1.028, and 1.047. Now, we find the irregular components by SI/S, for $1.042/1.028 = 1.014$, $1.043/1.047 = .996$, and so on, yielding .977, 1.014, .982, 1.007, 1.030, .993, .981, 1.010, .990, .992, 1.061, .924, 1.044, 1.001, 1.001, .994, .966, and 1.002.

12.15 To find the moving averages, we average out terms 12 at a time, yielding 25.717, 26.117, 26.597, 26.9, 27.467, 27.883, 28.663, 29.925, 29.742, 30.150, 30.733, 31.3, 31.617, 31.95, 32.483, 33.017, 33.675, 34.4, 34.983, 35.483, 36.05, 36.5, 36.992, 37.242, 37.633, 38.083, 38.35, 38.692, 39.333, 40.092, 40.75, 41.767, 42.25, 42.667, 43.017, 43.550, and 43.942. The centered averages are $(25.717+26.117)/2 = 25.917$, $(26.117+26.597)/2 = 26.357$, and so on, yielding 26.7485, 27.1835, and so on, yielding n=36, $\Sigma TC = 1250.2025$, and $\Sigma X(TC) = 25,113.723$. Therefore, we have $b = 0.5109335$, and $a = 25.276$, so that $T = 25.276 + 0.5109335X$. Substituting X=1,2,3,...,35,36 gives values for T, starting with 25.79, 26.30, ..., 43.16, 43.67.

To generate C, we divide each centered average by the T value, yielding $25.917/25.79 = 1.005$, $26.357/26.30 = 1.002$, and so on, yielding .998, .995, ..., 1.003, 1.002.

To generate SI, we divided Y by the TC term, with $30.2/25.917 = 1.231$, $31.9/26.357 = 1.191$, .927, .960, ..., 1.202, 1.227, and 1.237.

For the S terms, we averaged out every twelvth term, so that the terms add to twelve, yielding .793, .857, .830, 1.190, 1.182, 1.238, .914, .983, .886, and .740.

For the i terms, divide SI/S = $1.231/1.226 = 1.004$, $1.191/1.161 = 1.026$, 1.014, .977, .995, ..., 1.010, 1.038, and .999.

12.17 For the first quarter, 1984, we have $T = 300.85 + 20.179411(19) = 684.26$

$F = 684.26(1.00)(.855)(.95) = 555.79$, if conservative, and

$F = 684.26(1.00)(.855)(1.05) = 614.29$, if aggressive.

For the first quarter, 1985, we have $T = 300.85 + 20.179411(23) = 764.98$

$F = 764.98(1.00)(.855)(.95) = 621.35$, if conservative, and

$F = 764.98(1.00)(.855)(1.05) = 686.76$, if aggressive.

12.19 For the fourth quarter, 1986, we have $T = 850.1610 - 5.2644(26) = 713.2866$

$F = 713.2866(1.00)(1.208)(.95) = 818.568$, if conservative, and

$F = 713.2866(1.00)(1.208)(1.05) = 904.733$, if aggressive.

12.21 For June, 1987, $T = 25.276 + .5019335(48) = 49.801$

$F = 49.801(1)(1.238)(1.05) = 64.74$

For August, 1987, $T = 25.276 + .5019335(50) = 50.823$

$F = 50.823(1)(1.159)(1.05) = 61.85$

For November, 1987, $T = 25.276 + .5019335(53) = 52.335$

$F = 52.335(1)(.886)(1.05) = 48.69$

12.23 For Week 6, we have $F = (875+923+1027+843+929)/5 = 919.4$, for an error of $899 - 919.4 = -20.4$

For Week 7, we have $F = (923+1027+843+929+899)/5 = 924.2$, for an error of $735 - 924.2) = -189.2$

Similarly, for weeks 8-13, we have 886.6, 885.2, 911.6, 923.4, 925.0, and 913 for errors of 133.4, 89.8, 76.4, -24.4, -182, and 47.

12.25 For May, we have $F = .25(86) + .75(86) = 86$, for an error of $75 - 86 = -11$.

For June, we have $F = .25(75) + .75(86) = 83.25$, for an error of $70 - 83.25 = -13.25$.

For July, we have $F = .25(70) + .75(83.25) = 79.9375$, for an error of $82 - 79.9375 = 2.0625$

For August, we have $F = .25(82) + .75(79.9375) = 80.4531$, for an error of $90 - 80.4531 = 9.5469$

For September, we have $F = .25(90) + .75(80.4531) = 82.8398$, for an error of $69 - 82.8398 = -13.8398$

12.27 No. Models assume that the general trends identified continue indefinitely. This could have been a "miracle" or a rare event.

12.29 (a) For the fourth quarter, 1989, we take the averages of the previous four quarters, yielding $(105,550+105625+104450+105990)/4 = 105,403.75$

12.31 (a) MAD = (|40.7 − 39.3|+|35.8 − 37.3|+|44.6 − 44.8|+|47.3 − 50.6|+|50.1 − 52.1|)/5 = 8.4/5 = 1.68

MSE = [(40.7 − 39.3)^2+(35.8 − 37.3)^2+(44.6 − 44.8)^2+(47.3 − 50.6)^2+(50.1 − 52.1)^2]/5 = 19.14/5 = 3.828

(b) MAD = (|40.7 − 38.7|+|35.8 − 39.0|+|44.6 − 42.0|+|47.3 − 43.0|+|50.1 − 49.8|)/5 = 12.4/5 = 2.48

MSE = [(40.7 − 38.7)^2+(35.8 − 39.0)^2+(44.6 − 42.0)^2+(47.3 − 43.0)^2+(50.1 − 49.8)^2]/5 = 39.58/5 = 7.916

12.33

Y	F	|e|	e^2
7300	7199.12	100.88	10176.7744
7658	7703.65	45.65	2083.9225
8801	8607.02	193.98	37628.2404
8595	8713.47	118.47	14035.1409
		458.98	63924.0782

MAD = 458.98/4 = 114.745

MSE = 63924.0782/4 = 15981.02

12.35

Y	F	|e|	e^2
65.8	64.74	1.06	1.1236
63.7	61.96	1.74	3.0276
47.1	48.69	1.59	2.5281
		4.39	6.6793

MAD = 4.39/3 = 1.463

MSE = 6.6793/3 = 2.2264

12.37 MAD = [20.4+189.2+133.4+89.8+76.4+24.4+182+47]/8 = 762.6/8 = 95.325

12.39 (a) The moving averages are 303.25, 323.25, 337.75, 341.25, 354, 362.25, 378.5, 380.75, 396.25, 417.25, 423, 428, 437.5, 449.5, 456.5, 459.5, and 477.25. Centered moving averages are 313.25, 330.5, 339.5, 347.625, 358.125, 370.375, 379.625, 388.5, 406.75, 420.125, 425.5, 432.75, 443.5, 453, 458, and 468.375.

(b) With n=16, ΣTC=6335.50 and ΣX(TC) = 57368.5, we have b = 10.3434 and a = 308.050. Therefore, T = 308.05 + 10.343X. Substituting in X=1,2,...,16 yields 318.39, 328.74, 339.08, 349.42, 359.77, 370.11, 380.45, 390.80, 401.14, 411.48, 421.83, 432.17, 442.51, 452.86, 463.20, and 473.54.

(c) With C = TC/T, we have 313.25/318.39 = .984, 330.5/328.74 = 1.005, 1.001, .995, .995, 1.001, .998, .994, 1.014, 1.021, 1.009, 1.001, 1.002, 1.000, .989, and .989. There does not appear to be a major cyclical effect on the data.

(d) First we need to find the SI terms, by finding Y/TC = 131/313.25 = .418, 384/330.5 = 1.162, 1.328, 1.108, .405, 1.174, 1.275, 1.158, .379, 1.183, 1.335, 1.093, .392, 1.181, 1.345, and 1.070. To find the seasonal components, we average out every fourth term, omitting the high and low, yielding (1.328+1.335)/1.3315, (1.108+1.093)/2 = 1.1005, .3985, and 1.1775. As these add up to 4.008, we need to multiply each term by 4/4.008 = .998, yielding 1.329, 1.098, .398, 1.175. We expect a very small third quarter index.

(e) I = SI/S = .418/.398 = 1.05, 1.162/1.175 = .989, .999, 1.009, 1.018, .999, .959, 1.055, .952, 1.007, 1.005, .995, .985, 1.005, 1.012, and .974.

(f) For 1987, first quarter, X=23, and T = 308.05 + 10.3434(23) = 545.9482, and F = 545.9482(1.329)(.95) = 689.

For 1987, second quarter, X=24, and T = 308.05 + 10.3434(24) = 556.2916, and F = 556.2916(1.098)(.95) = 580.

For 1987, third quarter, X=25, and T = 308.05 + 10.3434(25) = 566.6350, and F = 566.6350(.398)(.95) = 214.

For 1987, fourth quarter, X=26, and T = 308.05 + 10.3434(26) = 576.9784, and F = 576.9784(1.175)(.95) = 644.

12.41 For each quarter, we eliminate the high and low values and average the other 5 values, yielding .7870, .8198, 1.5018, and .8720. As these add to 3.9806, we multiply each term by 4/3.9806 = 1.0049, yielding .791, .824, 1.509, and .876.

12.43 (a) Starting with May, we have $.2(.45) + .8(.45) = .45$

June: $.2(.51) + .8(.45) = .4620$

July: $.2(.45) + .8(.45) = .4756$

August: $.2(.45) + .8(.45) = .4745$

September: $.2(.45) + .8(.45) = .4976$

October: $.2(.45) + .8(.45) = .5181$

November: $.2(.45) + .8(.45) = .5005$

December: $.2(.5) + .8(.5005) = .5004$

(b) MAD = $[|.51 - .45| + |.53 - .4620| + ... + |.5 - .5005| + |.4 - .5004|]/8 = .5405/8 = .0676$

MSE = $[(.51 - .45)^2 + (.53 - .4620)^2 + ... + (.5 - .5005)^2 + (.4 - .5004)^2]/8 = .0499231/8 = .0062$

12.45 The moving averages are 23.5, 26.25, 29, 30.75, 32.5, 35.25, 37.75, 40, 41, 45, 47.25, 49, 50.25, 53.5, 56.5, 57.75, 59.25, 61.5, 63, 65, and 65.75. The centered moving averages are 24.875, 27.625, 29.875, ..., 62.25, 64, and 65.375. With n=20, ΣTC = 925.125, $\Sigma X(TC)$ = 11,161.50, b = 2.177, a = 23.40, so T = 23.40 + 2.177X. Substituting in X=1,2,...,19,20 yields T = 25.577, 27.754, ..., 64.763, and 66.94.

To find the C values, we have TC/T = 24.875/25.577 = .973, 27.625/27.754 = .995, .998, .985, ..., .988, and .977.

We have SI = Y/TC = 23/24.875 = .9246, 16/27.89 = .5792, 1.4059, ..., 1.4219, and 1.1013. To determine the S values, for each quarter, we eliminate the high and low values and average in the remaining three, so for the first quarter, we have (1.4059+1.4338+1.4219)/3 = 1.4205, and 1.1153, .8931, and .5925 for the other quarters. As these add to 4.0214, we need to multiply each term by 4/4.0214 = .9947, yielding 1.413, 1.109, .888, and .589.

To find the irregular components, we divide SI/S = .9246/.888 = 1.041, .5792/.589 = .983, ..., 1.006, and .993.

12.47 For this, we will divide the figures by the seasonal term to make the adjustment. Therefore, we have 6800/.5 = 13,600 for I, 1985, 24385/1.70 = 14,344.12 for II, 1985, 20280/1.5 = 13,520 for III, 1985, and 4700/.3 = 15,666.67 for IV, 1985. For 1986, we have 7700/.5 = 15,400, 23914/1.7 = 14,067.06, 24050/1.5 = 16,033.33, and 4650/.3 = 15,500 for quarters I to IV respectively.

12.49 (a) We generate the moving averages over 12 items, yielding 37.833, 38.417, 39.083, ..., 48.167, 49.083, and 49.667. The TC are 38.125, 38.75, 39.458, ..., 47.917, 48.625, and 49.375. This results in n=24, ΣTC = 1047.956, $\Sigma X(TC)$ = 13652.283, b = .4807, and a = 37.656. Therefore, we have T = 37.656 + 0.4807X. Substituting in X=1,2,...,23,24 gives T = 38.137, 38.637, ... , 48.712, and 49.193.

To find the C terms, we divide the TC/T, yielding 38.125/38.137 = 1.000, 38.75/38.617 = 1.003, 1.009, ..., .993, .998, 1.004.

To find the SI terms, we take Y/TC = 13/38.25 = .341, 16/38.75 = .413, 1.09, ..., .960, .617, and .203. For the S terms, we take the average of each month's terms, yielding (1.306+1.300)/2 = 1.303 for January, (1.275+1.190)/2 = 1.233 for February, and 1.144, .960, .603, .159, .340, .418, 1.064, 1.373, 1.592, and 1.831 for March through December. Multiplying each term by the adjustment factor, 12/12.02 = .998 yields 1.300, 1.231, 1.142, .958, .602, .159, .339, .417, 1.062, 1.370, 1.589, 1.827.

For the irregular terms, we have SI/S = .341/.339 = 1.006, .413/.417 = .990, ... , 1.025, and 1.277.

(b) To deseasonalize the data, we divide each term by the S factor, yielding $46/1.300 = 35.38$ for January, 1984, $44/1.231 = 35.74$, 34.15, ..., 51.09, 52.86, and 52.00.

(c) For February, 1987, X=32, so $T = 37.656 + 0.4807(32) = 53.0384$. We have
$F = 53.0384(1.231)(.95) = 62$, being conservative, and
$F = 53.0384(1.231)(1.05) = 69$, being aggressive.
For August, 1987, X=38, so $T = 37.656 + 0.4807(38) = 55.9226$. We have
$F = 55.9226(.417)(.95) = 22$, being conservative, and
$F = 55.9226(.417)(1.05) = 24$, being aggressive.

12.51 (a) We generate the moving averages over 12 items, yielding 24.167, 25.5, 26.583, ..., 74.417, 75.333, and 77.5. The TC terms are 24.834, 26.042, 27.083, ..., 73.875, 74.875, and 76.417. This results in n=26, $\Sigma TC = 1810.342$, $\Sigma X(TC) = 39218.377$, $b = 1.4741$, and $a = 23.016$. Therefore, we have $T = 23.016 + 1.4741X$. Substituting in X=1,2,...,35,36 gives $T = 24.49, 25.964, 27.438, ..., 73.135, 74.610$, and 76.084.

To find the C terms, we divide the TC/T, yielding $24.834/24.49 = 1.014$, $26.042/25.964 = 1.003$, .987, ..., 1.010, 1.004, and 1.004.

To find the SI terms, we take $Y/TC = 33/24.834 = 1.329$, $41/26.042 = 1.574$, .923, ..., .880, 1.028, and 1.269. For the S terms, we take the average of each month's terms, yielding $(.897+.861+.939)/3 = .899$, for January, $(.703+.716+.667)/3 = .695$ for February, and .684, .850, 1.017, 1.320, 1.364, 1.516, .901, .837, .694, and 1.231 for March through December. Multiplying each term by the adjustment factor, $12/12.008 = .999$ yields .898, .694, .683, .849, 1.016, 1.319, 1.363, 1.515, .900, .836, .693, and 1.230. For the irregular terms, we have $SI/S = 1.329/1.363 = .975$, $1.574/1.515 = 1.040$, 1.026, ..., 1.037, 1.012, and .962.

(b) For January, 1987, we have X=43, so $T = 23.016 + 1.4741(43) = 86.4023$, so $F = 86.4023(.898)(1.05) = 81$
For June, 1987, we have X=48, so $T = 23.016 + 1.4741(43) = 93.7728$, so $F = 93.7728(1.319)(1.05) = 130$
For August, 1987, we have X=50, so $T = 23.016 + 1.4741(50) = 96.7210$, so $F = 96.7210(1.515)(1.05) = 154$
For October, 1987, we have X=52, so $T = 23.016 + 1.4741(52) = 99.6692$, so $F = 99.6692(.836)(1.05) = 87$

(c) $\text{MAD} = [|85 - 81| + |113 - 130| + |148 - 154| + |94 - 87|]/4 = 34/4 = 8.5$
$\text{MSE} = [(85 - 81)^2 + (113 - 130)^2 + (148 - 154)^2 + (94 - 87)^2]/4 = 390/4 = 97.5$

12.53 They might affect it greatly. If they can be accurately forecast, we should include them as independent variables in the trend model, and make individual forecasting models for each.

CHAPTER 13

INDEX NUMBERS

13.1 For revenue passengers, we need to divide each term by 748,000. Therefore, we have $100(679/748) = 90.8$, $75400/748 = 100.8$, on so on, yielding 100, 100.5, 102.3, 106.6, 113.1, 119.1, 127, 142.2, and 150.8 for years 1980 − 1988.

For Operating revenues, we need to divide each term by 87,676. Therefore, we have $100(58769/87676) = 67.0$, $7075500/87676 = 80.7$, on so on, yielding 100, 106.1, 106.3, 112.1, 119.5, 128, 140.3, 159.9, and 172.4 for years 1980 − 1988.

For operating expenses, we need to divide each term by 88310. Therefore, we have $100(55669/88310) = 63.0$, $7001900/88310 = 79.3$, on so on, yielding 100, 106.1, 105.8, 108.9, 112.9, 122.4, 134.2, 152, and 165.8 for years 1980 − 1988.

13.3 (a) The value index for 1985 is $\dfrac{1700(1.20) + 10{,}850(.58) + 60{,}010(2.50)}{1165(1.15) + 8450(.70) + 65{,}315(3.25)} = \dfrac{158{,}358}{219{,}528.5}(100) = 72.1$

(b) The Paasche index for 1985 is $\dfrac{1700(1.20) + 10{,}850(.58) + 60{,}010(2.50)}{1700(1.15) + 10{,}850(.70) + 60{,}010(3.25)} = \dfrac{158{,}358}{204{,}582.5}(100) = 77.4$

(c) The Laspeyres index for 1985 is $\dfrac{1165(1.20) + 8{,}450(.58) + 65{,}315(2.50)}{1165(1.15) + 8450(.70) + 65{,}315(3.25)} = \dfrac{169{,}586.5}{219{,}528.5}(100) = 77.3$

(d) The fixed-weights price index for 1985 is $\dfrac{1020(1.20) + 7870(.58) + 50{,}400(2.50)}{1020(1.15) + 7870(.70) + 50{,}400(3.25)}$
$= \dfrac{131{,}788.6}{170{,}482}(100) = 77.3$

(e) The Paasche quantity index for 1985 is $\dfrac{1700(1.20) + 10{,}850(.58) + 60{,}010(2.50)}{1165(1.20) + 8{,}450(.58) + 65{,}315(2.50)}$
$= \dfrac{158{,}358}{169{,}586.5}(100) = 93.4$

(f) The Laspeyres quantity index for 1985 is

$\dfrac{1700(1.15) + 10850(.70) + 60{,}010(3.25)}{1165(1.15) + 8450(.70) + 65{,}315(3.25)} = \dfrac{204{,}582.5}{219{,}528.5}(100) = 93.2$

(g) The fixed weight quantity index for 1985 is

$$\frac{1700(1.60) + 10850(.65) + 60{,}010(3.10)}{1165(1.60) + 8450(.65) + 65{,}315(3.10)} = \frac{195{,}803.5}{209{,}833}(100) = 93.3$$

13.5 (a) The Paasche index for 1987 is $\frac{4(1.05) + 6(.95) + 4(.79) + 5(.65)}{4(.98) + 6(.89) + 4(.60) + 5(.58)} = \frac{16.31}{14.56}(100) = 112.0$

The Laspeyres index for 1987 is $\frac{2(1.05) + 6(.95) + 5(.79) + 3(.65)}{2(.98) + 6(.89) + 5(.60) + 3(.58)} = \frac{13.7}{12.04}(100) = 113.8$

The fixed-weights price index for 1987 is $\frac{3(1.05) + 6(.95) + 7(.79) + 2(.65)}{3(.98) + 6(.89) + 7(.60) + 2(.58)}$
$= \frac{15.68}{13.64}(100) = 115.0$

(b) The Paasche quantity index for 1987 is $\frac{4(1.05) + 6(.95) + 4(.79) + 5(.65)}{2(1.05) + 6(.95) + 5(.79) + 3(.65)}$
$= \frac{16.31}{13.7}(100) = 119.1$

The Laspeyres quantity index for 1987 is
$\frac{4(.98) + 6(.89) + 4(.60) + 5(.58)}{2(.98) + 6(.89) + 5(.60) + 3(.58)} = \frac{14.56}{12.04}(100) = 120.9$

The fixed weight-quantity index for 1987 is
$\frac{4(1.05) + 6(.80) + 4(.70) + 5(.60)}{2(1.05) + 6(.80) + 5(.70) + 3(.60)} = \frac{14.8}{12.2}(100) = 121.3$

(c) The value index for 1986 is $\frac{3(1.05) + 6(.80) + 7(.70) + 2(.60)}{2(.98) + 6(.89) + 5(.60) + 3(.58)} = \frac{14.05}{12.04}(100) = 116.7$

The value index for 1987 is $\frac{4(1.05) + 6(.95) + 4(.79) + 5(.65)}{2(.98) + 6(.89) + 5(.60) + 3(.58)} = \frac{16.31}{12.04}(100) = 135.5$

13.7 Yes. If the decrease in meat consumption is significantly less than the proportional increase in chicken and fish consumption, then this can, in fact, happen. For this to happen, basically, the increase would have to less than the absolute difference in price between the red meat and chicken, or meat/fish.

If the price of fish has increased, but is still less than the price of red meat, the bill could actually decrease, or not necessarily increase

13.9 (a) Year 1 is the base year, with an index of 100.

(b) For year 0, we have $[(100 - 89)/89]100 = 12.36\%$

For year 1, we have $(6/100)100 = 6\%$

For year 2, we have $(9/115)100 = 8.49\%$

For year 3, we have $(2/117)100 = 1.71\%$

(c) We have $586(117/100) = \$685.62$

(d) For year 3, we have $(115/106)(3.89) = \$4.22$

For year 4, we have $(116/106)(3.89) = \$4.29$

13.11 (a) The base year is 1982.

(b) We would have $(94/75)(900) = \$1,128$

(c) For 1980, we have $[(150 - 230)/230](100) = 34.8\%$

For 1981, we have $[(100 - 150)/150](100) = 33.3\%$

For 1982, we have $[(94 - 100)/100](100) = 6\%$

For 1983, we have $[(87 - 94)/94](100) = 7.4\%$

For 1984, we have $[(76 - 87)/87](100) = 12.6\%$

For 1985, we have $[(75 - 76)/76](100) = 1.33\%$

For 1986, we have $[(62 - 75)/75](100) = 17.3\%$

13.13 (a) We have $(115/183)300,000 = \$188,524.59$

(b) $(260/130)300,000 = 600,000$

(c) $(80/153)300,000 = 156,862.74$

(d) $(80/90)300,000 = 266,666.67$

(e) $(147/250)300,000 = 176,400$

(f) $(275/118)300,000 = 699,152.54$

13.15 For 1977, we have $(100/105)5,370 = 5,114.29$

For 1978, we have $(100/113)6,845 = 6,057.52$

For 1979, we have $(100/122)7,421 = 6,082.79$

For 1980, we have $(100/138)8,650 = 6,268.12$

The increase can be attributed to an actual increase in value plus the effects of inflation.

13.17 (a) It means that in real dollars, the average cost for these areas would cause a rejection of the H_0: average area cost = average national cost.

(b) Letting New England be the base requires that we divide each term by $122/100 = 1.22$. Therefore, we have 100 for New England, $101/1.22 = 82.79$ for Middle Atlantic, $99/1.22 = 81.15$ for South Atlantic, and 77.87, 79.51, 80.33, 77.05, 80.33, and 85.25 for the other areas, respectively.

13.19 (a) The base year is 1961.

(b) We need to divide each term by 125 to make 1965 the base year. Therefore, we have $(97/125)100 = 77.6$ for 1960, $(100/125)100 = 80$ for 1961, and 85.6, 88.8, 96.8, 100, 103.2, 108.0, and 112 for 1962 – 1968.

(c) We need to divide each term by 49 to make 1965 the base year. Therefore, we have $(97/49)100 = 198.0$ for 1960, $(100/49)100 = 204.1$ for 1961, and 218.4, 226.5, 246.9, 255.1, 263.3, 275.5, and 285.7 for 1962 – 1968.

13.21 We need to divide each term by 119 to change the base year to year 6. Therefore, we have $(77/119)100 = 64.7$, $(89/119)100 = 74.8$ for year 2, and 84.0, 89.9, 96.6, and 100 for years 3 – 6.

13.23 (a) We have $(100/97)100 = 103.1$ for 1961, $(107/100)100 = 107$ for 1962, $(111/107)100 = 103.7$ for 1963, and 109.0, 103.3, 103.2, 104.7, and 103.7 for 1964 – 1968.

(b) $I_{68,61} = \dfrac{103.7(104.7)...(107.0)}{100^6} = 140$, the same as the 1968 index with 1961 as the base year.

(c) $I_{68,65} = \dfrac{103.7(104.7)(103.2)(103.3)}{100^2} = 112$, the same as the 1968 index with 1965 as the base year.

13.25 We have $(89/77)100 = 115.6$ for year 2, $(100/89)100 = 112.4$ for year 3, and 107.0, 107.5, and 103.5 for years 4 – 6.

(a) $I_{6,4} = \dfrac{103.5(107.5)}{100} = 111.3$

(b) $I_{6,3} = \dfrac{103.5(107.5)(107)}{100^2} = 119.1$

(c) $I_{5,2} = \dfrac{(107.5)(107)(112.4)}{100^2} = 129.3$

(d) $I_{5,1} = \dfrac{(107.5)(107)(112.4)(115.6)}{100^3} = 149.5$

(e) $I_{4,2} = \dfrac{(107)(112.4)}{100} = 120.3$

(f) $I_{4,1} = \dfrac{(107)(112.4)(115.6)}{100^2} = 139.0$

13.27 (a) The base year is 1986, as its adjusted figure is the same as the actual expenditure.

(b) If we call actual expenditure X and adjusted expenditures Y, we have $\Sigma X = 15.05$, $\Sigma Y = 17.40$, $\Sigma X^2 = 32.97$, $\Sigma Y^2 = 43.94$, and $\Sigma XY = 37.84$, so that $SS_{xy} = 0.43$, $SS_x = 0.61$, and $SS_y = .6854$. Therefore,

$r = \dfrac{0.43}{\sqrt{0.61(0.6854)}} = .665$

If you change X and Y, we can say that $(.665)^2 = .44$ or 44% of the variation in expenditures is directly attributed to inflation.

(c) For this to happen, then the inflation rate was higher than the percentage increase in actual expenditures.

13.29 (a) For revenue, using 1984 as a base year, we divide each term by 12.1 and multiply by 100, yielding $(14.3/12.1)100 = 118.2$ for 1987, $(14/12.1)100 = 115.7$ for 1986, 114.9 for 1985, and 100 for 1985.

For net income, using 1984 as a base year, we divide each term by 185.5 and multiply by 100, yielding $(154.3/185.5)100 = 83.2$ for 1987, $(144.5/185.5)100 = 77.9$ for 1986, 83.3 for 1985, and 100 for 1985.

(b) Obviously, since 1984, costs have increased at a much higher pace than revenues. Therefore, we have a decrease in net income.

13.31 The advantage of the Laspeyres over the Paasche is that as the current year changes, weights do not. Comparisons between years are possible. The advantage of fixed-weights over Laspeyres is that we can be sure that the weights are representative.

13.33 (a) For 1973, we have $1.45(\$2,185,000) = \$3,168,250$. For 1975, we have $1.70(\$2,185,000) = \$3,714,500$.

(b) To shift the base to 1972, we divide each index by 127 and multiply by 100. We have 100 for 1972, $(145/127)100 = 114.2$ for 1973, and 143.3, 133.9, and 148.8 for $1974 - 1976$.

13.35 A simple index deals with only one product or commodity; an aggregate index deals with more than one.

13.37 For 1980, we divide each term by 75 and multiply by 100. Therefore, we have 100 for 1980, $(97/75)100 = 129.3$ for 1981, and 133.3, 138.7, 149.3, 153.3, and 146.7 for $1982 - 1986$.

Since sales in 1975 were 80% of 1980, then we divide each term in the 1980 index by 80 and multiply by 100, for $(100/80)100 = 125$ for 1980, $(129.3/80)100 = 161.6$ for 1981, and 166.6, 173.4, 186.6, 191.6, and 183.4 for $1982 - 1986$.

13.39 An economic indicator is any time series that is related to the average activity in the economy or part of it. If the terms in the series are percentages, the economic indicator is an index.

13.41 Unweighted price and quantity index numbers create two problems: (1) the unit of measure affects the index, and (2) each product contributes equally regardless of its importance.

13.43 (a) For price, 1982 is the base year. For Quantity, 1979 is the base year.

(b) To shift the price index to 1980, we divide each term by $(104/100) = 1.04$. Therefore, we have $120/1.04 = 115.4$ for 1978, $111/1.04 = 106.7$ for 1979, and 100, 92.3, 96.2, 91.3, 85.6, 102.9, 110.6, and 113.5 for 1980 – 1987.

To shift the quantity index to 1980, we divide each term by $(97/100) = .97$. Therefore, we have $115/.97 = 118.6$ for 1978, $100/.97 = 103.1$ for 1979, and 100, 82.5, 106.2, 110.3, 118.6, 130.9, 149.5, and 168.0 for 1980 – 1987.

13.45 An absolute indicator is a time series with actual quantities. A relative indicator is a percentage or a relative term.

CHAPTER 14

DECISION MAKING

14.1 (a) The actions are to prepare 10, 11, 12, 13, or 14 tortes.

(b) Demand is 10, 11, 12, 13, or 14 tortes.

(c)

Demand	Number Prepared				
	10	11	12	13	14
10	260	248	236	224	212
11	260	286	274	262	250
12	260	286	312	300	288
13	260	286	312	338	326
14	260	286	312	338	364

(d)

Demand	Number Prepared				
	10	11	12	13	14
10	0	12	24	36	48
11	26	0	12	24	36
12	52	26	0	12	24
13	78	52	26	0	12
14	104	78	52	26	0

14.3 (a) Accept shipment to Kansas City or accept shipment to Salt Lake City

(b) The return shipment from Salt Lake City is or is not available.

(c)

SLC Return	Shipment Accepted	
	KC	SLC
Yes	1800	3200
No	1800	900

(d)

SLC Return	Shipment Accepted	
	KC	SLC
Yes	1400	0
No	0	900

14.5 (a) The options are to try to rent at 450 per unit, try to rent at 500 per unit, or sublease at 800 for the two units.

(b) The payoff is affected by whether the market rent is at 450 or 500.

(c)

	Rent		
Results	450	500	Lease
Both rent	900	1000	800
1 rents	450	500	800
0 rent	0	0	800

(d)

	Rent		
Results	450	500	Lease
Both rent	100	0	200
1 rents	350	300	0
0 rent	800	800	0

14.7 (a)

Precipitation	Snow	French Drains
Heavy	73,800	−15,200
Moderate	67,800	59,800
Light	18,000	92,800

Precipitation	Snow	French Drains
Heavy	0	89,000
Moderate	0	8,000
Light	91,000	0

(b) For the second year, we do not have the start up costs, so we just increase our monthly projections by 7%. This yields

Precipitation	Snow	French Drains
Heavy	95,016	−7,704
Moderate	88,596	72,546
Light	17,976	107,856

Precipitation	Snow	French Drains
Heavy	0	102,700
Moderate	0	16,050
Light	89,880	0

(c) Because doing so would not consider all states of nature which include such states as heavy snow first year and light snow second year. There are 9 possible states (HH, HM, HL, MH, MM, ML, LH, LM, LL).

14.9 (a) For maximax strategy, we choose B since the max profit is $7,000,000.

(b) For maximin strategy, we choose D, since the min is 2,500. The other mins are 0, −1300, and −500.

(c)

Snowfall	Resort A	B	C	D
0 – 5	2500	3800	3000	0
5 – 10	1550	3250	2650	0
10 – 15	300	2300	0	1550
15 +	2000	0	1500	4250

Using the minimax regret strategy, we choose Resort A, since its worst case of 2500 is less than the 3800, 3000, and 4250 of B, C, and D, respectively.

14.11 For maximax strategy, hire, since $300,000 > 0$.

For maximin strategy, hire, since both worst cases are $-120,000$, but the other possibilities for hire are all positive.

For minimax regret strategy, we need to make an opportunity loss table. For hire, the values are 120,000, 0, 0, 0, so the worst case is 120,000. For don't hire, the values are 0, 30,000, 120,000, and 420,000, so 420,000 is the worst case. Since $120,000 < 420,000$, then hire.

14.13 (a) For the maximax strategy, we accept Salt Lake City, with the max value of 3200.

(b) For the maximin strategy, we accept Kansas City since $1800 > 900$.

(c) For the minimax regret strategy, we accept Salt Lake City since $900 < 1400$.

14.15 For the maximax strategy, prepare 14, with the highest payoff of 364.

For the maximin strategy, prepare 10, with the highest min payoff of 260.

For the minimax regret strategy, prepare 13, with the highest OL of 36.

14.17 (a) For snow removal, we have $73,800(.50) + 67,800(.27) + 1800(.23) = 55,620$

For French drains, we have $-15,200(.50) + 59,800(.27) + 92,800(.23) = 29,890$

(b) For snow removal, we have $0(.50) + 0(.27) + 91,000(.23) = 20,930$

For French drains, we have $89,000(.50) + 8,000(.27) + 0(.23) = 46,660$

(c) Snow removal has a higher expected payoff and a lower expected regret.

(d) Perfect information would be knowing the snowfall in advance.

(e) For perfect information, the expected payoff would be $73,800(.50) + 67,800(.27) + 92,800(.23) = 76,550$.

(f) EVPI $= 76,550 - 55,620 = 20,930$.

14.19 (a) For A, the expected payoff $= 0(.05) + 1200(.40) + 4000(.25) + 5000(.30) = 2980$.

For B, the expected payoff $= -1300(.05) + -500(.40) + 2000(.25) + 7000(.30) = 2335$.

Similarly, the expected payoffs are 2740 and 2737.5 for C and D.

(b) For A, the expected opportunity loss = $2500(.05) + 1550(.40) + 300(.25) + 2000(.30) = 1420$. For B, the expected opportunity loss = $3800(.05) + 3250(.40) + 2300(.25) + 0(.30) = 2065$. Similarly, the expected opportunity losses are 1660 and 1662.5 for C and D.

(c) For perfect information, the expected payoff = $2500(.05) + 2750(.40) + 4300(.25) + 7000(.30) = 4400$. Therefore, EVPI $= 4400 - 2980 = 1420$.

14.21 (a) The expected payoff for hire is $-120,000(.25) + 25,000(.20) + 100,000(.20) + 300,000(.35) = 100,000$. For don't hire, the expected payoff is $0(.25) + -5,000(.20) + -20,000(.20) + -120,000(.35) = -47,000$. It makes sense to hire.

(b) The expected regret for hire is $120,000(.25) + 0(.20) + 0(.20) + 0(.35) = 30,000$. For don't hire, the expected regret is $0(.25) + 30,000(.20) + 120,000(.20) + 420,000(.35) = 177,000$. It makes sense to hire.

(c) The expected payoff for perfect information is $0(.25) + 25,000(.20) + 100,000(.20) + 300,000(.35) = 130,000$. Therefore, EVPI $= 130,000 - 100,000 = 30,000$.

14.23 We rewrite our payoff table to look as below, with the probabilities given in parentheses:

	Rent		
Results	450	500	Lease
Both rent	900(.9025)	1000(.64)	800
1 rents	450(.0950)	500(.32)	800
0 rent	0(.0025)	0(.04)	800
(a) E(Payoff)	855	800	800

Therefore, rent at 450.

(b) We rewrite our opportunity loss table to look as below, with the probabilities given in parentheses:

	Rent		
Results	450	500	Lease
Both rent	100(.9025)	0(.64)	200
1 rents	350(.0950)	300(.32)	0
0 rent	800(.0025)	800(.04)	0
E(OL)	125.5	128	128 or 180.5, depending on which probabilities are used

Therefore, rent at 450.

(c) The expected payoff using perfect information is $1000(.64) + 800(.32) + 800(.04) = 928$.

(d) EVPI $= 928 - 855 = 73$.

14.25

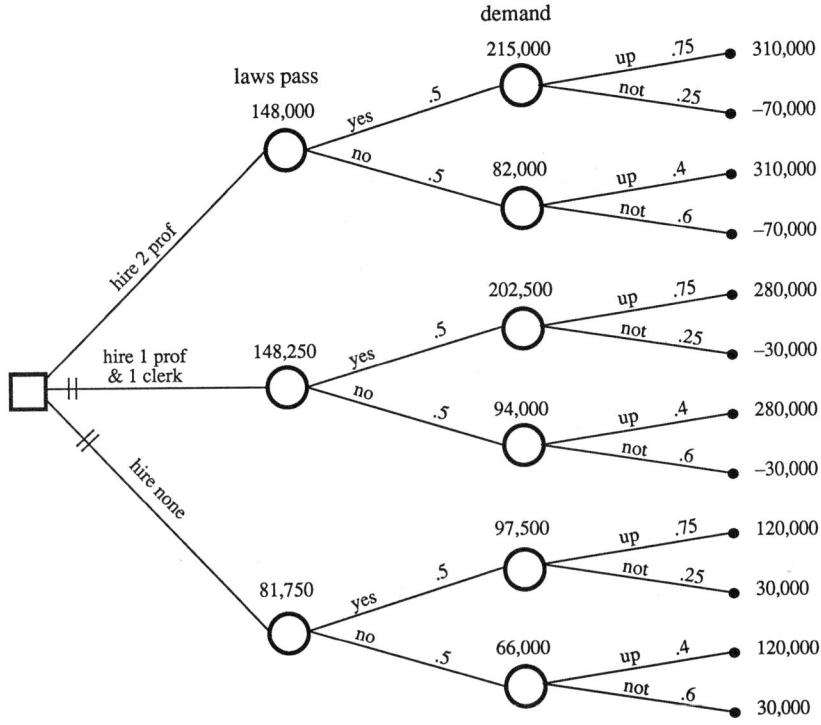

Hire two professionals. The expected payoff is 148,500.

14.27

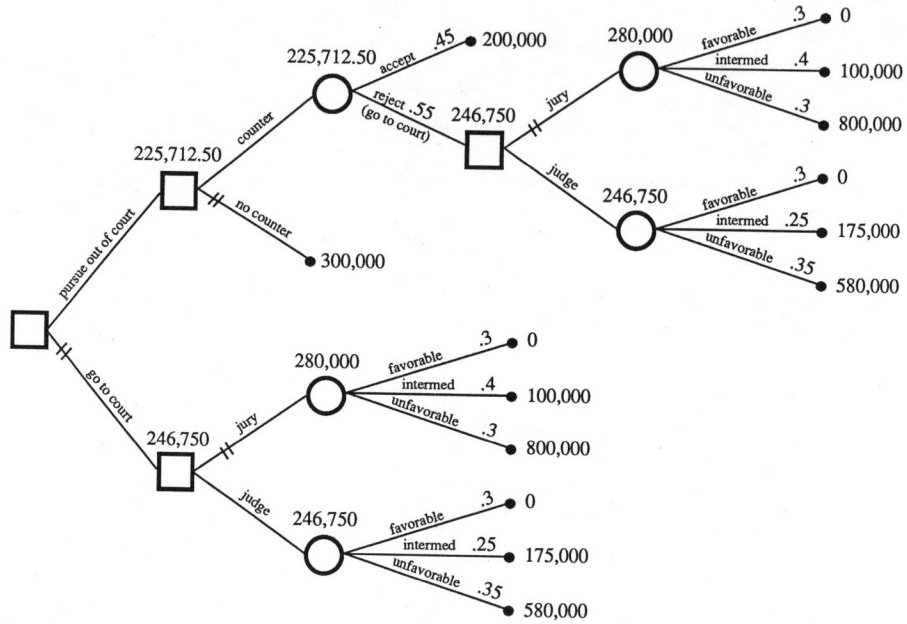

The insurance company should pursue out of court, but take the judge if forced into court. The expected payoff for out of court settlement is $225,712.50.

14.29 (a) For 195 people showing up, this is 5 below capacity, so the loss is 5(250) = 1250. For 196, the loss is 4(250) = 1000, and 750, 500, 250, and 0 for 197 through 200. For 201, this requires 1 bump, so the loss is 400, 800 for 202, and 1200, 1600, and 2000 for 203 − 205.

(b) With n=205, and p=.95, the actual probabilities for 195 − 205 are .128, .124, .108, .083, .055, .031, .015, .006, .002, .000, and .000.

(c) To find the expected loss = ΣLP(L), we first need to adjust the probabilities, since they do not add up to 1. Since They add up to .552, we divide each term by .552, so E(L) = [1250(.128) + 1000(.124) + ... + 2000(.000)]/.552 = 785.

(d) When the number of people who show up is more than the 200, the loss is 400(number who show − 200). When the number of people who show up is less than 200, then the loss is 250(200 − number who show).

(e) We leave this as an exercise. There are just too many branches to put in the manual.

(f) When the number of people with reservations who show is ≤ 199, this causes new branches for the number of standbys.

(g) It will affect E(P) or E(Loss). Therefore, it has an effect on the strategy.

14.31 For the Ltd. Layer Option, in the first circle, we have $94(.3) + 93.5(.7) = 93.65$ For the second circle, we have $94(.3) + 93.5(.7) = 93.15$, and 92.65, 92.15, 91.65, and 91.15 for the other circles. For the thinner layers, we have $93.65(.6) + 93.15(.4) = 93.45$ for high, 92.45 for medium, and 91.45 for low. Therefore, the overall expected value is $93.45(.2) + 92.45(.5) + 91.45(.3) = 92.35$.

14.33 (a) The expected payoffs are \$23,625 for accept and 18,000 for reject.

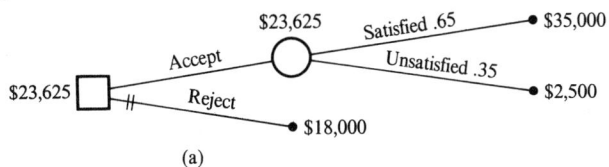

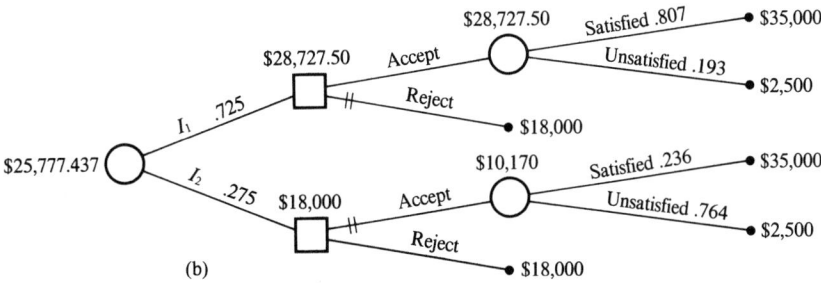

(b) If we let I_1 be the event that the test says satisfactory and I_2 be the event that the test says unsatisfactory, then

$P(I_1) = P(I_1|S)P(S) + P(I_1|U)P(U) = .90(.65) + .40(.35) = .725$

$P(I_2) = P(I_2|S)P(S) + P(I_2|U)P(U) = .10(.65) + .60(.35) = .275$

$P(S|I_1) = \dfrac{P(S)P(I_1|S)}{P(S)P(I_1|S) + P(U)P(I_1|U)} = \dfrac{.65(.90)}{.65(.90) + .35(.40)} = .807$

$P(U|I_1) = \dfrac{P(U)P(I_1|U)}{P(S)P(I_1|S) + P(U)P(I_1|U)} = \dfrac{.35(.40)}{.65(.90) + .35(.40)} = .193$

$P(S|I_2) = \dfrac{P(S)P(I_2|S)}{P(S)P(I_2|S) + P(U)P(I_2|U)} = \dfrac{.65(.10)}{.65(.10) + .35(.60)} = .236$

$P(U|I_2) = \dfrac{P(U)P(I_2|U)}{P(S)P(I_2|S) + P(U)P(I_2|U)} = \dfrac{.35(.60)}{.65(.10) + .35(.60)} = .764$

The expected payoff is \$25,777.44

(c) The most the supervisor should pay is $\$25,777.44 - 23,625 = \$2152.44$

14.35 (a) If the survey is not conducted, the expected first year payoff is $.43(\$2,000,000) + .39(0) + .18(-1,000,000) = \$680,000$.

(b)

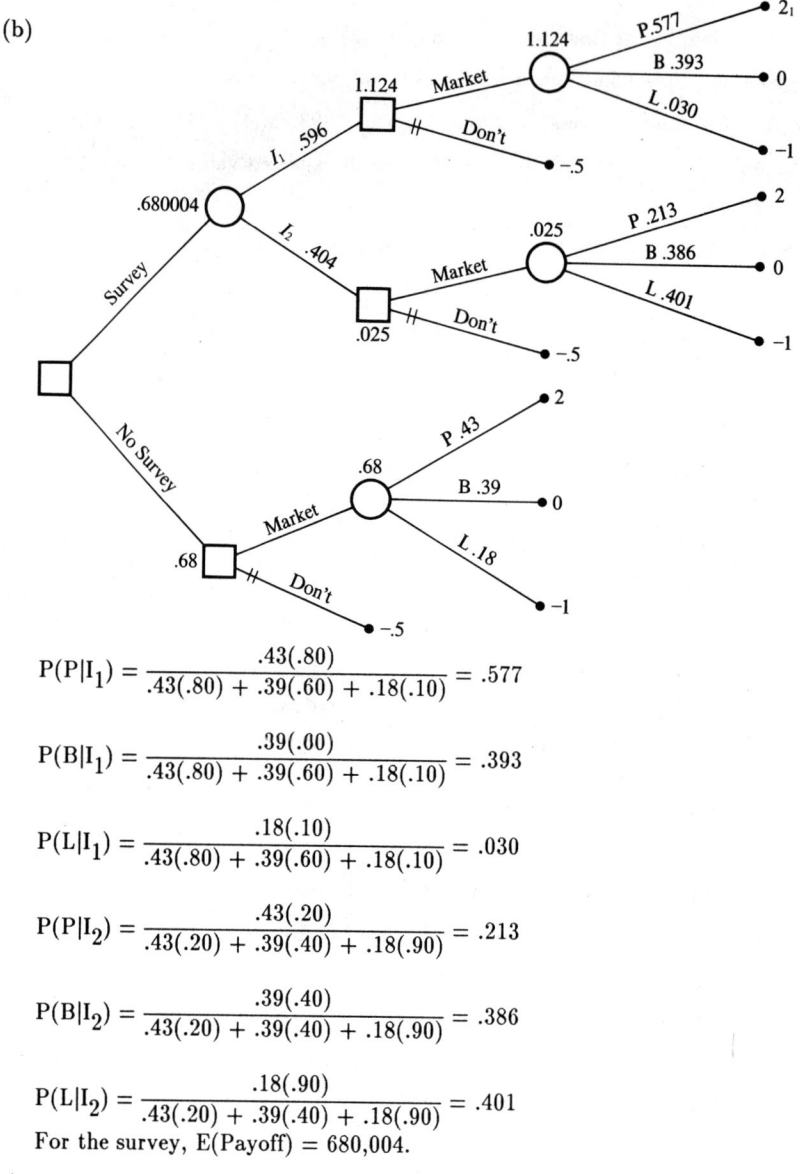

$$P(P|I_1) = \frac{.43(.80)}{.43(.80) + .39(.60) + .18(.10)} = .577$$

$$P(B|I_1) = \frac{.39(.00)}{.43(.80) + .39(.60) + .18(.10)} = .393$$

$$P(L|I_1) = \frac{.18(.10)}{.43(.80) + .39(.60) + .18(.10)} = .030$$

$$P(P|I_2) = \frac{.43(.20)}{.43(.20) + .39(.40) + .18(.90)} = .213$$

$$P(B|I_2) = \frac{.39(.40)}{.43(.20) + .39(.40) + .18(.90)} = .386$$

$$P(L|I_2) = \frac{.18(.90)}{.43(.20) + .39(.40) + .18(.90)} = .401$$

For the survey, E(Payoff) = 680,004.

(c) 680,004 − 680,000 = $4

14.37 (a) If we do not purchase the insurance, the expected cost is 1000(.80) + 8000(.15) + 15,000(.05) = $2,750. The expected cost of the policy is 3000 + 2000(.05) = 3100. The expected cost of the insurance policy is higher, so the expected payoff is − $2,750.

− 110 −

(b)

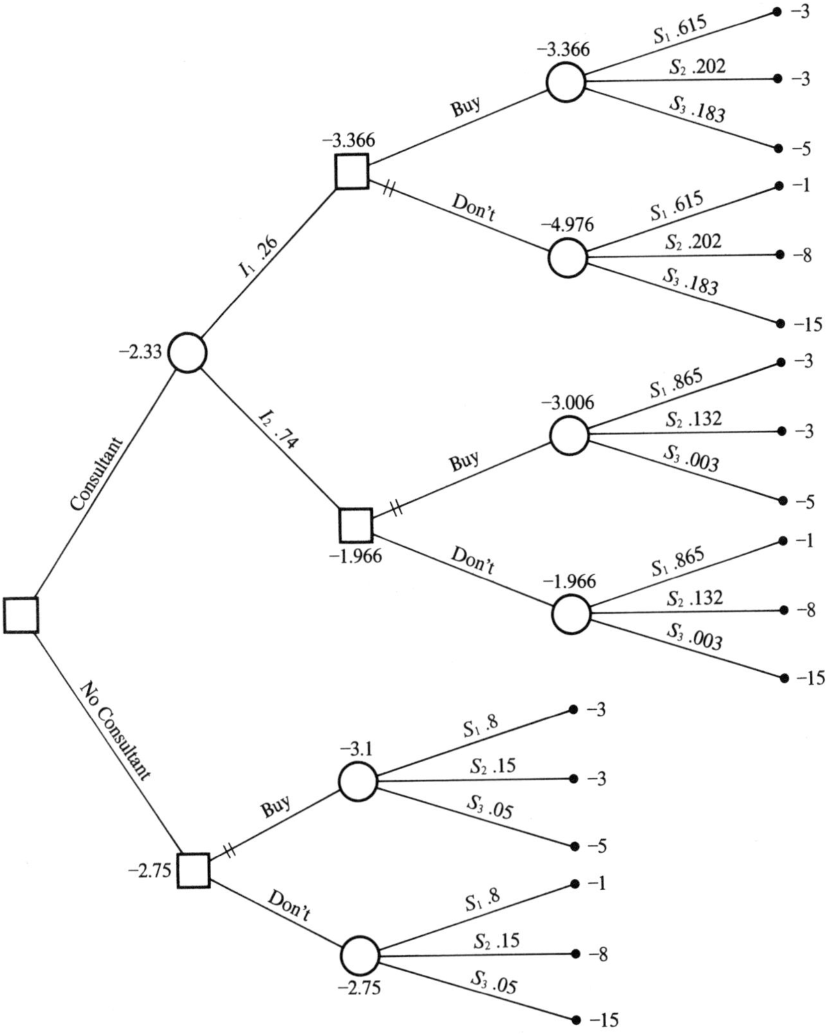

Consultant
No Consultant
−2.33
I_1 .26
I_2 .74
−3.366
Buy
Don't
−3.366
−4.976
S_1 .615
S_2 .202
S_3 .183
−3
−3
−5
−1
−8
−15
−1.966
−3.006
−1.966
S_1 .865
S_2 .132
S_3 .003
−3
−3
−5
−1
−8
−15
−2.75
−3.1
−2.75
S_1 .8
S_2 .15
S_3 .05
−3
−3
−5
−1
−8
−15

$$P(S_1|I_1) = \frac{.8(.2)}{.8(.2) + .15(.35) + .05(.95)} = .615$$

$$P(S_2|I_1) = \frac{.15(.35)}{.8(.2) + .15(.35) + .05(.95)} = .202$$

$$P(S_3|I_1) = \frac{.05(.95)}{.8(.2) + .15(.35) + .05(.95)} = .183$$

$$P(S_1|I_2) = \frac{.8(.8)}{.8(.8) + .15(.65) + .05(.05)} = .865$$

$$P(S_2|I_2) = \frac{.15(.65)}{.8(.8) + .15(.65) + .05(.05)} = .132$$

$$P(S_3|I_2) = \frac{.05(.05)}{.8(.8) + .15(.65) + .05(.05)} = .003$$

Yes, they should buy the insurance. The expected payoff given I_1 is $-3,366$, which is less than the expected cost of the insurance policy, so we should purchase the insurance.

(c) Given I_2, the expected payoff is $-1,966$, which is better than the expected cost of the policy, so don't purchase the policy.

(d) If the consultant's report is used, the expected payoff is $-2,330$.

(e) The value of the report is $2750 - 2330 = \$420$.

14.39 The maximax decision is to order 650, since it has the max payoff of 3240. The maximin decision is to order 400, since its min of 2000 is larger than any other min. The minimax regret strategy says to order 650, to minimize the worst case opportunity loss at 375.

14.41 (a) For expected payoffs, for prepare 10, $E(P) = 260$. For 11, we have $248(.12) + 286(.23) + 286(.14) + 286(.35) + 286(.16) = 281.44$. For 12, we have $236(.12) + 274(.23) + 312(.14) + 312(.35) + 312(.16) = 294.14$, 301.52 for 13, and 295.6 for 14. For expected opportunity loss, for 10, we have $0(.12) + 26(.23) + 52(.14) + 78(.35) + 104(.16) = 57.2$. For 11, we have $12(.12) + 0(.23) + 26(.14) + 52(.35) + 78(.16) = 35.76$. For 12, we have 23.06, 15.68 for 13, and 21.6 for 14. Preparing 13 gives the highest expected payoff and smallest expected opportunity loss.

(b) EVPI $= 260(.12) + 286(.23) + 312(.14) + 338(.35) + 364(.16) = 317.2$.

14.43 (a) The maximax strategy is to rent at 500, since the payoff of 1000 is the maximum.

(b) The maximin decision would be to lease to the management company for a "no-risk" 800 return.

(c) The minimax-regret decision would be to lease to the management company since the opportunity loss max of 200 would be best "worst case".

14.45 (a)

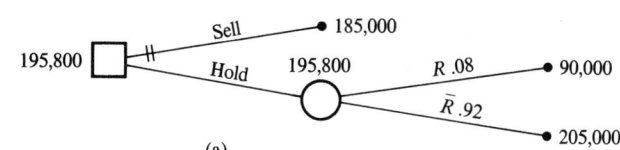

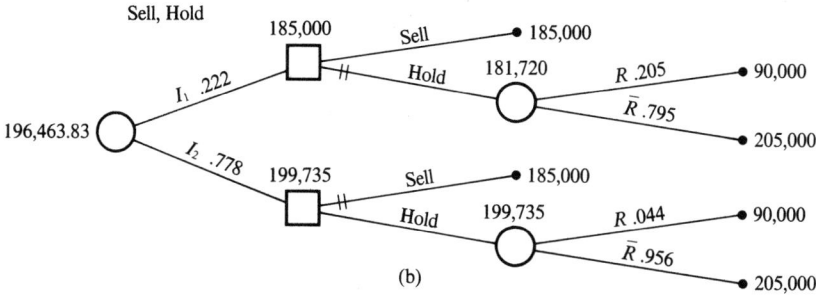

Hold. E(P) = 205,000(.92) + 90,000(.08) = 195,800 > 185,000

(b) If we let I_1 = a recession is predicted, and I_2 = a recession is not predicted, then

$$P(R|I_1) = \frac{.08(4/7)}{.08(4/7) + .92(10/52)} = .205$$

$$P(\overline{R}|I_1) = \frac{.92(10/52)}{.08(4/7) + .92(10/52)} = .795$$

$$P(R|I_2) = \frac{.08(3/7)}{.08(3/7) + .92(42/52)} = .044$$

$$P(\overline{R}|I_2) = \frac{.92(42/52)}{.08(3/7) + .92(42/52)} = .956$$

If a recession is predicted, sell. E(P) = 185,000. If a recession is not predicted, then hold. E(P) = 199,735.

(c) From the consultant's report, the expected payoff is 196,463.83.

(d) Without the consultant, the expected payoff was 195,800. 196,463.83 − 195,800 = 663.83

14.47 (a) To maximize expected payoff, for KC we have 1800, while for SLC, we have 3200(.35) + 900(.65) = 1705. Therefore, accept the Kansas City shipment.

(b) For KC, E(OL) = 1400(.35) = 490. For SLC, E(OL) = 900(.65) = 585. Accept the KC shipment.

(c) To maximize expected payoff, for KC we have 1800, while for SLC, we have 3200(.85) + 900(.15) = 2855. Therefore, accept the Salt Lake City shipment. For KC, E(OL) = 1400(.85) = 1190. For SLC, E(OL) = 900(.15) = 135. Accept the SLC shipment.

14.49 (a) For both years 1 and 2, maximax is French Drain, with the maximum payoffs of 112,000 and 128,400. For maximin strategy, choose Snow Removal for both years, as 21,000 and 38,520 are better than the worst cases of 4,000 and 12,840.

(b) For the minimax regret decisions, for Year 1, we choose French drains, since $89,000 < 91,000$. For Year 2, we choose Snow Removal, since $89,880 < 102,720$.

14.51 (a) Accept a one year lease, since $-\$17,000$ is the "maximum" payoff.

(b) Accept a two year lease, since $-22,000 > -30,000$.

(c) Accept a two year lease, since $5000 < 8000$.

(d) For a one year lease, $E(P) = -17,000(.67) + -30,000(.33) = -21,290 > 22,000$. Therefore, accept a one year lease.

(e) For a one year lease, $E(OL) = 8,000(.33) = 2,640$. For a two year lease, $E(OL) = 5,000(.67) = 3350$. Therefore, accept a one year lease.

(f) Given perfect information, $E(P) = -17,000(.67) + -22,000(.33) = -18,650$.

(g) EVPI $= -18,650 - (21,290) = 2,640$.

14.53 (a)

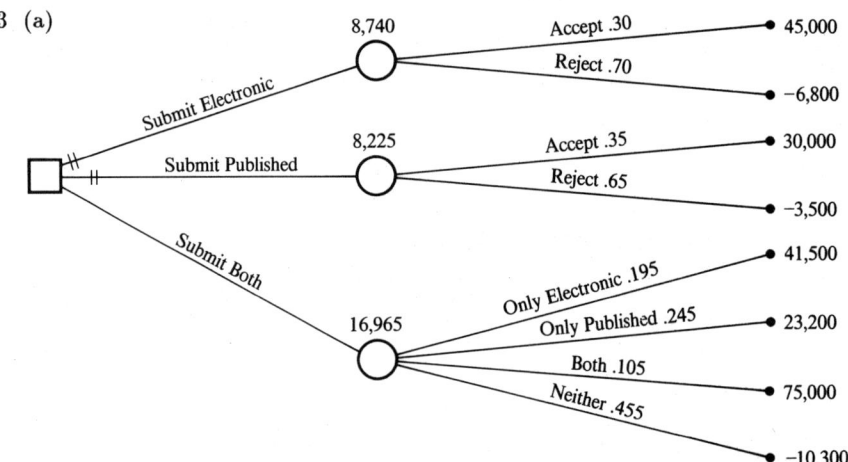

Since the events are independent, if we submit both, $P(E \cap \overline{P}) = .30(.65) = .195$, $P(\overline{E} \cap P) = .70(.35) = .245$, $P(E \cap P) = .30(.35) = .105$, $P(\overline{E} \cap \overline{P}) = .70(.65) = .455$.

(b) Submit both. The expected payoff is 16,965.

14.55 (a) For order 30, $E(P) = 22.50$. For 31, $E(P) = 21.75(.08) + 23.25(.10) + \ldots + 23.25(.02) = 23.13$. For 32, $E(P) = 21.00(.08) + 22.50(.10) + \ldots + 24.00(.02) = 23.61$, 23.79 for 33, 23.45 for 34, 22.73 for 35. Therefore, we choose to order 33.

(b) For order 30, $E(P) = 0(.08) + .75(.10) + \ldots + 3.75(.02) = 1.99$. For 31, $E(P) = .75(.08) + 0(.10) + \ldots + 3.00(.02) = 1.36$, .88 for 32, .70 for 33, 1.04 for 34, 1.76 for 35. Therefore, we choose to order 33.

(c) $E(PPI) = 22.50(.08) + 23.25(.10) + \ldots + 26.25(.02) = 24.49$

(d) Perfect information would be knowing the demand ahead of time or a consignment agreement.

(e) EVPI = 24.49 − 23.79 = .70 per week.

CHAPTER 15

CHI-SQUARE TESTS

15.1 The means of the distributions are the degrees of freedom. Therefore, we have (a) 8 (b) 17 (c) 21 (d) 30 (e) 40 (f) 50

15.3 For H_1: $\sigma^2<9000$, we have $\chi^2 = 29(8550)/9000 = 27.55 > \chi^{2*} = 17.7$. Do not reject H_0 at $\alpha=.05$. No, there is insufficient evidence to conclude that the sale can be completed.

15.5 For H_1: $\sigma^2<200$, we have $s^2 = 159.36$ and $\chi^2 = 7(159.36)/200 = 5.58 > \chi^{2*} = 2.83$. Do not reject H_0 at $\alpha=.10$. No, it cannot be concluded that the new schedule has been effective in reducing variance.

15.7 For H_0: $p_1=.5$, $p_2 = .34$, $p_3 = .09$, and $p_4 = .07$, our expected totals are generated by multiplying the fractions by the sample size, 500, yielding 250, 170, 45, and 35. Therefore, we have

$$\chi^2 = \frac{(293-250)^2}{250} + \frac{(140-170)^2}{170} + \frac{(39-45)^2}{45} + \frac{(28-35)^2}{35} = 14.89 > \chi^{2*}_{.01,3} = 11.3.$$

Reject H_0 at $\alpha=.01$. The distribution has changed since 1988.

15.9 To test for an equal percentage in each category, we have H_0: $p_1 = p_2 = p_3 = p_4 = p_5 = .2$. We expect $260(.2) = 52$ in each group, so

$$\chi^2 = \frac{(89-52)^2}{52} + \frac{(39-52)^2}{52} + \ldots + \frac{(38-52)^2}{52} = 34.46 > \chi^{2*} = 7.78, \text{ based on 4 df.}$$

Reject H_0 at $\alpha=.10$. Yes, there is not an equal percentage of members in each category.

15.11 (a) For H_1: $p \neq .68$, we have $z = \dfrac{.80-.68}{\sqrt{.68(.32)/50}} = 1.82 < z^* = 1.96$.

Do not reject H_0 at $\alpha=.05$. No, there is not sufficient evidence to conclude that the Washington students differ significantly from the overall population.

(b) We have H_0: $p_1=.275$, $p_2=.096$, $p_3=.135$, $p_4=.021$, $p_5=.262$, $p_6=.211$, for the expected group totals, we multiply each fraction by 50, yielding 13.75, 4.8, 6.75, 1.05, 13.1, and 10.55. With the small counts in the second and fourth groups, we combine those with the third group to make the categories nothing, less than \$1000, \$1000 or more, don't know.
$$\chi^2 = \frac{(10-13.75)^2}{13.75} + \frac{(25-12.60)^2}{12.60} + \frac{(5-13.1)^2}{13.1} + \frac{(10-10.55)^2}{10.55} = 18.26 > \chi^{2*} = 7.81,$$
based on 3 df. Reject H_0 at $\alpha=.05$. Yes, the distributions of amounts differs for the Washington students.

15.13 We have H_0: $p_1 = .45$, $p_2 = .35$, $p_3 = .2$. To find the expected totals, we multiply each value by n=500, yielding 225, 175, and 100. Therefore, we have
$$\chi^2 = \frac{(250-225)^2}{225} + \frac{(167-175)^2}{175} + \frac{(83-100)^2}{100} = 6.04 > \chi^{2*} = 5.99, \text{ based on 2 degrees}$$
of freedom. Reject H_0 at $\alpha=.05$. Yes, the figures refute the congressional report.

15.15 We have H_0: $p_1=.19$, $p_2=.19$, $p_3=.19$, $p_4=.19$, $p_5 = .24$. With n=200, the expected group totals are 38, 38, 38, 38, and 48. Therefore, we have
$$\chi^2 = \frac{(48-38)^2}{38} + \frac{(67-38)^2}{38} + \frac{(32-38)^2}{38} + \frac{(23-38)^2}{38} + \frac{(30-48)^2}{48} = 38.38 > \chi^{2*} =$$
9.49, based on 4 degrees of freedom. Reject H_0 at $\alpha=.05$. Yes, there has been a significant change in the group support.

15.17 (a) These are the 11 leading sodas. There are others whose market share is less than 1.6%.

(b) With the smallest groups having 1.6%, we need $n(.016) \geq 5$, so $n \geq 5/.016 = 312.5$ or 313 people. We would have 12 categories(1 for "others") and therefore 11 degrees of freedom.

15.19 To test H_0: The distribution is approximately normal with $\mu=50$, $\sigma=10$, we have

Terms	z	Prob in Interval	E	O
< 40	< −1	.1587	23.81	20
40 − 45	−1 to −0.5	.1498	22.47	27
45 − 50	−0.5 to 0	.1915	28.73	32
50 − 55	0 to 0.5	.1915	28.73	27
55 − 60	0.5 to 1	.1498	22.47	21
> 60	> 2	.1587	23.81	23

We have
$$\chi^2 = \frac{(20-23.81)^2}{23.81} + \frac{(27-22.47)^2}{22.47} + \ldots + \frac{(23-23.81)^2}{23.81} = 2.12 < \chi^{2*} = 11.1 \text{ based on 5}$$
df. Do not reject H_0 at $\alpha=.05$. There is insufficient evidence to contradict a normal distribution with $\mu=50$ and $\sigma=10$.

15.21 We have

Lifetime	z	Prob in Interval	E	O
< 4.5	< −0.71	.2388	15.76	13
4.5 – 5.0	−0.71 to 0	.2612	17.24	12
5.0 – 5.5	0 to 0.71	.2612	17.24	15
5.5 – 6.0	0.71 to 1.43	.1624	10.72	9
6.0 – 6.5	1.43 to 2.14	.0602	3.97	6
> 6.5	> 2.14	.0162	1.07	11

We need to combine the bottom two groups, since the expected group sizes are less than 5. Now, for H_0: the data forms a normal distribution with a $\mu=5$, $\sigma=.7$,

$$\chi^2 = \frac{(13-15.76)^2}{15.76} + \frac{(12-17.24)^2}{17.24} + \cdots + \frac{(17-5.04)^2}{5.04} = 31.01 > \chi^{2*} = 9.49, \text{ based on 4}$$

degrees of freedom. Reject H_0 at $\alpha=.05$. The data does not form a normal distribution with $\mu=5$ and $\sigma=.7$.

15.23 (a) The df becomes $6-1-2 = 3$, so $\chi^{2*} = 7.81$.

(b) Using 4.25 and 6.75 as midpoints of the outer classes, we have $\overline{X} = 5.37$, and $s = 0.86$. Now we have

Lifetime	z	Prob in Interval	E	O
< 4.5	< −1.01	.1562	10.31	13
4.5 – 5.0	−1.01 to −0.43	.1774	11.71	12
5.0 – 5.5	−0.43 to 0.15	.2260	14.92	15
5.5 – 6.0	0.15 to 0.73	.2077	13.71	9
6.0 – 6.5	0.73 to 1.31	.1376	9.08	6
> 6.5	> 1.31	.0951	6.28	11

$$\chi^2 = \frac{(13-10.31)^2}{10.31} + \frac{(12-11.71)^2}{11.71} + \cdots + \frac{(11-6.28)^2}{6.28} = 6.92 < 7.81.$$

There is insufficient evidence at $\alpha=.05$ to reject the hypothesis of a normal distribution.

15.25 (a) $P(X=0) = \frac{1.5^0 e^{-1.5}}{0!} = .223$, $P(X=1) = \frac{1.5^1 e^{-1.5}}{1!} = .335$, $P(X=2) = \frac{1.5^2 e^{-1.5}}{2!} = .251$, $P(X=3) = \frac{1.5^3 e^{-1.5}}{3!} = .126$, $P(X=4) = \frac{1.5^4 e^{-1.5}}{4!} = .047$, $P(X=5) = \frac{1.5^5 e^{-1.5}}{5!} = .014$, $P(X=6) = \frac{1.5^6 e^{-1.5}}{6!} = .004$

(b) To find the expected frequencies, we multiply each probability by 80, yielding 17.84, 26.80, 20.08, 10.08, 3.76, 1,12, and .32. With the small expected totals For X=4,5, and 6, we group these into "4 or more." To test H_0: the distribution is Poisson, we have

$$\chi^2 = \frac{(18-17.84)^2}{17.84} + \frac{(29-26.80)^2}{26.80} + \ldots + \frac{(3-5.20)^2}{5.20} = 1.271 < \chi^{2*} = 9.49, \text{ based on } 4$$

df. Do not reject H_0 at $\alpha=.05$. There is insufficient evidence to disprove that the sample came from a Poisson distribution with $\lambda=1.5$.

(c) If λ is unknown, we estimate it with $\overline{X} = [0(18) + 1(29) + \ldots + 6(1)]/90 = 112/90 = 1.4$, but we lose an additional df from estimating λ. We have

$$P(X=0) = \frac{1.4^0 e^{-1.4}}{0!} = .247, \; P(X=1) = \frac{1.4^1 e^{-1.4}}{1!} = .345, \; P(X=2) = \frac{1.4^2 e^{-1.4}}{2!} =$$

$.242, \; P(X=3) = \frac{1.4^3 e^{-1.4}}{3!} = .113, \; P(X=4) = \frac{1.4^4 e^{-1.4}}{4!} = .039, \; P(X=5) =$

$\frac{1.4^5 e^{-1.4}}{5!} = .011, \; P(X=6) = \frac{1.4^6 e^{-1.4}}{6!} = .003$

To find the expected frequencies, we multiply each probability by 80, yielding 19.76, 27.6, 19.36, and 13.28(we combine X=3,4,5,6,... due to small cell expectancies). To test H_0: the distribution is Poisson, we have

$$\chi^2 = \frac{(18-19.76)^2}{19.76} + \frac{(29-27.6)^2}{27.6} + \ldots + \frac{(12-13.28)^2}{13.28} = 0.49 < \chi^{2*} = 5.99, \text{ based on } 2$$

df. Do not reject H_0 at $\alpha=.05$. There is insufficient evidence to disprove that the sample came from a Poisson distribution with $\lambda=1.4$.

15.27 (a) The row totals are 16, 20, 68, 39, 7, and 1. The column totals are 20, 65, 4, 16, and 2, for an overall total of 151. This creates an abundance of cells with small expected values, so we consolidate the data for Pre-Policy categories as < 1 and ≥ 1, and Post-Policy as $< 1/2$, $1/2 - 1$, and ≥ 1, creating the table below, with expected totals in parentheses:

Post-Policy	Pre-Policy < 1	Pre-Policy ≥ 1	Totals
< 1/2	31(20.26)	5(15.74)	36
1/2 − 1	46(38.28)	22(29.72)	68
≥ 1	8(26.46)	39(20.54)	47
	85	66	151

To test H_1: Post-Policy results are related to Pre-Policy results,

$$\chi^2 = \frac{(31-20.96)^2}{20.96} + \frac{(46-38.28)^2}{38.28} + \ldots + \frac{(39-20.54)^2}{20.54} = 46.03 > \chi^{2*} = 5.99 \text{ based on 2}$$

df. Reject H_0 at $\alpha=.05$. Post-Policy results are related to Pre-Policy results. Looking at the table, the numbers of people in the ≥ 1 group has been reduced, so the policy has been effective.

(b) [see (a)]

15.29 For H_1: Existence of a CMP is related to Sales, the row totals are 81, 72, and 13, while the column totals are 92 and 74. Therefore, the expected cell totals are $81(92)/166 = 44.89$, $72(92)/166 = 39.90$, 7.20, 36.11, 32.10, and 5.80.

$$\chi^2 = \frac{(28-44.89)^2}{44.89} + \frac{(52-39.90)^2}{39.90} + \ldots + \frac{(1-5.80)^2}{5.80} = 29.64 > \chi^{2*} = 5.99 \text{ based on 2}$$

df. Reject H_0 at $\alpha=.05$. Existence of a crisis management plan is related to sales.

For H_1: Existence of a CMP is related to Number of Employees, the row totals are 104, 53, and 26, while the column totals are 92 and 91. Therefore, the expected cell totals are $104(92)/183 = 52.28$, $53(92)/183 = 26.64$, 13.07. 51.72, 26.36, and 12.93.

$$\chi^2 = \frac{(47-52.28)^2}{52.28} + \frac{(31-26.64)^2}{26.64} + \ldots + \frac{(12-12.93)^2}{12.93} = 2.64 < \chi^{2*} = 5.99 \text{ based on 2}$$

df. Do not reject H_0 at $\alpha=.05$. Existence of a crisis management plan is not significantly related to number of employees.

15.31 To test H_0: Resolution is related to the presence/absence of an arbitrator, we make the following table, with expected totals in parentheses.

	Settled	Not Settled	Totals
Arbitration	38(32.4)	12(17.6)	50
No Arbitration	43(48.6)	32(26.4)	75
	81	44	125

$$\chi^2 = \frac{(38-32.4)^2}{32.4} + \frac{(43-48.6)^2}{48.6} + \frac{(12-17.6)^2}{17.6} + \frac{(32-26.4)^2}{26.4} = 4.59 > \chi^{2*} = 3.84,$$

based on 1 df. Reject H_0 at $\alpha=.05$. Yes, resolution is related to the presence/absence of an arbitrator.

15.33 The row totals are 96, 53, and 26, while column totals are 37, 77, and 61. Expected cell totals are $96(37)/175 = 20.3$, $53(37)/175 = 11.21$, ..., 18.47, 9.06. For H_1: Opinions differ among Groups, we have

$$\chi^2 = \frac{(18-20.3)^2}{20.3} + \frac{(10-11.21)^2}{11.21} + \ldots + \frac{(18-18.47)^2}{18.47} + \frac{(10-9.06)^2}{9.06} = 4.74 < \chi^{2*} =$$

9.49, based on 4 degrees of freedom. Do not reject H_0 at $\alpha=.05$. There is no evidence of differing opinions among locations.

15.35 (a) The row totals are 314 and 320, while the column totals are 123, 149, 195, 141, and 26 for an overall total of 634. The expected cell counts are $314(123)/634 = 60.92$, $320(123)/634 = 62.08$, ..., 12.88, 13.12. For H_1: Insurance Coverage is related to Education, we have

$$\chi^2 = \frac{(38-60.92)^2}{60.92} + \frac{(85-62.08)^2}{62.08} + \ldots + \frac{(15-12.88)^2}{12.88} + \frac{(11-13.12)^2}{13.12} = 50.09 > \chi^{2*} =$$

13.3 based on 4 degrees of freedom. Reject H_0 at $\alpha=.01$. The relationship is established.

(b) We have H_0: $p_1 = .2$, $p_2 = .2$, $p_3 = .3$, $p_4 = .25$, $p_5 = .05$. With n=634, the expected group totals are 126.8, 126.8, 190.2, 158.5, and 31.7. We have

$$\chi^2 = \frac{(123-126.8)^2}{126.8} + \frac{(149-126.8)^2}{126.8} + \ldots + \frac{(26-31.7)^2}{31.7} = 7.08 < \chi^{2*} = 9.49, \text{ based on}$$

4 degrees of freedom. Do not reject H_0 at $\alpha=.05$. There is not sufficient evidence to contradict the percentage distribution in H_0.

15.37 The row totals are 54, 35, 34, and 27. The column totals are 54, 62, and 34 for an overall total of 150. This gives expected cell totals of $54(54)/150 = 19.44$, $35(54)/150 = 12.60$, ..., 7.71, 6.12. For H_1: Salary is related to Gender/Marriage, we have

$$\chi^2 = \frac{(20-19.44)^2}{19.44} + \frac{(13-12.60)^2}{12.60} + \ldots + \frac{(7-7.71)^2}{7.71} + \frac{(6-6.12)^2}{6.12} = 0.42 < \chi^{2*} = 10.6,$$

based on 6 degrees of freedom. Do not reject H_0 at $\alpha=.10$. There is not evidence of discrimination at this time.

15.39 The row totals are 16.0, 27.8, 20.3, and 12.9(million), while column totals are 45.1, 24.4, and 7.5, for a grand total of 77. Therefore, expected cell totals are $16(45.1)/77 = 9.37$, $27.8(45.1)/77 = 16.28$, ..., 1.98, and 1.26. These are all in millions, so there is no problem with expected cell sizes. For H_1: Education is related to Income, we have

$$\chi^2 = \frac{(13.5 - 9.37)^2}{9.37} + \frac{(17.1 - 16.28)^2}{16.28} + \cdots + \frac{(2.2 - 1.98)^2}{1.98} + \frac{(3.3 - 1.26)^2}{1.26} = 10.95 \text{ million}$$

$> \chi^{2*} = 12.6$, based on 6 degrees of freedom. Reject H_0 at $\alpha = .05$. Level is related to income.

15.41 We use this data to make the following table, with expected totals in parentheses:

	No Debt	Debt	Totals
< $10,000	67(35.13)	33(64.87)	100
10,000 – 19,999	48(35.13)	52(64.87)	100
20,000 – 34,999	62(70.25)	138(129.75)	200
35,000 – 49,999	69(105.37)	231(194.63)	300
$\geq$ 50,000	35(35.13)	65(64.87)	100
	281	519	800

For H_1: Debt is related to Income, we have

$$\chi^2 = \frac{(67 - 35.13)^2}{35.13} + \frac{(48 - 35.13)^2}{35.13} + \cdots + \frac{(231 - 194.63)^2}{194.63} + \frac{(65 - 64.87)^2}{64.87} = 72.71$$

15.43 For H_1: $\sigma \neq 1.5$, we have $s^2 = 5.07$. Therefore, $\chi^2 = 7(5.07)/2.25 = 15.78 > \chi^{2*} = 12.0$, based on 7 df. Reject H_0 at $\alpha = .20$. The machine should be shut down for repair. A type-one error involves shutting down the machine for repair, when there is not a problem. A type-two error involves letting the machine run when, in fact, the variation is too high and it is out of control.

15.45 We use the data to create the following table, with expected cell totals in parentheses.

	Phone	No Phone	Totals
Need Repair	12(15)	18(15)	30
Don't	38(35)	32(35)	70
	50	50	100

For H_1: Incidence of Repair is related to the presence of a phone number,

$$\chi^2 = \frac{(12-15)^2}{15} + \frac{(38-35)^2}{35} + \frac{(18-15)^2}{15} + \frac{(32-35)^2}{35} = 1.72 < \chi^{2*} = 3.84 \text{ based on 1}$$

df. Do not reject H_0 at $\alpha=.05$. There is not evidence that incidence of repair is related to the presence of a phone number.

15.47 (a) $\chi^2_{.01,\ 15} = 30.6$ (b) $\chi^2_{.975,\ 8} = 2.18$, $\chi^2_{.025,\ 8} = 17.5$ (c) $\chi^2_{.90,\ 17} = 10.1$
(d) $\chi^2_{.95,\ 25} = 14.6$, $\chi^2_{.05,\ 25} = 37.7$ (e) $\chi^2_{.05,\ 4} = 9.49$ (f) $\chi^2_{.10,\ 6} = 10.6$
(g) $\chi^2_{.01,\ 24} = 43.0$ (h) $\chi^2_{.05,\ 20} = 31.4$ (i) $\chi^2_{.05,\ 9} = 16.9$ (j) $\chi^2_{.10,\ 100} = 118.5$

15.49 The row totals are 73, 68, and 54. The column totals are 42, 56, 37, and 60, for an overall total of 195. The expected cell totals are $73(42)/195 = 15.72$, $68(42)/195 = 14.65$, ..., 20.92, 16.62. For H_1: Response is related to Economic Level,

$$\chi^2 = \frac{(7-15.72)^2}{15.72} + \frac{(30-14.65)^2}{14.65} + \ldots + \frac{(12-20.92)^2}{20.92} + \frac{(19-16.62)^2}{16.62} = 39.09 > \chi^{2*} =$$

12.6, based on 6 df. Reject H_0 at $\alpha=.05$. Yes, the variables are related.

15.51 (a) $\chi^2_{.05,9} = 16.9$ (b) $\chi^2_{.05,14} = 23.7$ (c) $\chi^2_{.05,24} = 36.4$

15.53 χ^2 tests are for <u>count</u> data, when we see how many observations fit into a category. We test the distribution of the <u>fractions</u> of responses for each category. In ANOVA, we deal with <u>measurement</u> data, where each observation represents a single result or observation. In the F test, we test the means of these observations.

15.55 $\chi^2_{.975,\ 15} = 6.26$

15.57 For H_0: the distribution is normal with $\mu=30$, $\sigma=6$, we create the following table:

Interval	z-scores	Prob	E	O
< 19.5	< −1.75	.0401	1.72	2
19.5 − 23.5	−1.75 to −1.08	.1000	4.30	7
23.5 − 27.5	−1.08 to −0.42	.1971	8.48	8
27.5 − 31.5	−0.42 to 0.25	.2615	11.24	15
31.5 − 35.5	0.25 to 0.92	.2225	9.57	4
35.5 − 39.5	0.92 to 1.58	.1217	5.23	6
> 39.5	> 1.58	.0571	2.46	1

We merge the top two categories together to generate expected group counts of at least five. Similarly, because the bottom group has an E value that is too small, it is merged "up." We have

$$\chi^2 = \frac{(9-6.02)^2}{6.02} + \frac{(8-8.48)^2}{8.48} + \frac{(15-11.24)^2}{11.24} + \frac{(4-9.57)^2}{9.57} + \frac{(7-7.69)^2}{7.69} = 6.05 < \chi^{2*}$$

$=7.78$, based on $5-1=4$ df. Do not reject H_0 at $\alpha=.10$. No, the normal distribution cannot be disputed at this time.

15.59 The row totals are 70 and 80, while the column totals are 20, 60, and 70. Therefore, the expected totals are $70(20)/150 = 9.33$, $80(20)/150 = 10.67$, 28, 32, 32.67, and 37.33. To test H_1: Scores differ between universities,

$$\chi^2 = \frac{(12-9.33)^2}{9.33} + \frac{(8-10.67)^2}{10.67} + \ldots + \frac{(47-37.33)^2}{37.33} = 10.07 > \chi^{2*} = 5.99, \text{ based on 2}$$

df. Reject H_0 at $\alpha=.05$. There is a difference between universities.

15.61 Both the independence and goodness-of-fit tests are one-tailed tests. Both tests determine if the χ^2 statistic is large enough to reject H_0. It becomes larger as differences between observed and expected values become larger.

15.63 Row totals are 100 for each group. Column totals are 140, 181, 87, and 92, for an overall total of 500. For each cell, expected totals are $100(T)/500$, where T is the column total. Therefore, we have 28 for each cell in Savings, 36.2 for Checking, 17.4 for Loan Repayment, and 18.4 for Bill Paying. For H_1: Service is related to Location,

$$\chi^2 = \frac{(34-28)^2}{28} + \frac{(22-28)^2}{28} + \ldots + \frac{(20-18.4)^2}{18.4} + \frac{(14-18.4)^2}{18.4} = 20.661 < \chi^{2*} = 21.0,$$

based on 12 df. Do not reject H_0 at $\alpha=.05$. No, there is insufficient evidence to conclude that the service utilized is related to the facility location.

15.65 Compare the distributions among the groups to see where there are differences.

15.67 For H_0: $p_1 = p_2 = p_3 = p_5 = .25$, we expect $75(.25) = 18.75$ in each group.

$$\chi^2 = \frac{(21-18.75)^2}{18.75} + \frac{(16-18.75)^2}{18.75} + \frac{(13-18.75)^2}{18.75} + \frac{(25-18.75)^2}{18.75} = 4.52 < \chi^{2*} = 7.81,$$

based on 3 df. Do not reject H_0 at $\alpha=.05$. No, there is insufficient evidence of a difference in preference.

15.69 To test H_0: Checks have a normal distribution with $\mu=\$35.60$, $\sigma=5.80$, we have the table which follows:

Check Total	z	P(z)	E	O
< 25.00	< −1.83	.0336	3.06	11
25.00 − 30.00	−1.83 to −0.97	.1324	12.05	35
30.00 − 35.00	−0.97 to −0.10	.2924	26.77	16
35.00 − 40.00	−0.10 to 0.75	.3132	28.50	20
> 40.00	> 0.75	.2266	20.62	9

We consolidate the top category into the second group, since the expected count of the < 25.00 group is too small. We have

$$\chi^2 = \frac{(46-15.11)^2}{15.11} + \ldots + \frac{(9-20.62)^2}{20.62} = 76.60 > \chi^{2*} = 7.81, \text{ based on 3 df.}$$

Reject H_0 at $\alpha=.05$. Yes, there has been a change in the check distribution at Juanita's.

15.71 For H_0: $p_1 = .30$, $p_2 = .25$, $p_3 = .15$, $p_4 = .30$, our expected group totals are 60, 50, 30, and 60. We have

$$\chi^2 = \frac{(43-60)^2}{60} + \frac{(32-50)^2}{50} + \frac{(68-30)^2}{30} + \frac{(57-60)^2}{60} = 59.58 > \chi^{2*} = 6.25, \text{ based on 3}$$

df. Reject H_0 at $\alpha=.10$. The distribution has changed in the past three years.

15.73 For H_0: $p_1 = p_2 = p_3 = p_4 = p_5 = .2$, we expect $.2(100) = 20$ for each group.

$$\chi^2 = \frac{(35-20)^2}{20} + \frac{(22-20)^2}{20} + \frac{(10-20)^2}{20} + \frac{(15-20)^2}{20} + \frac{(18-20)^2}{20} = 17.90 > \chi^{2*} =$$

13.3, based on 4 df. Reject H_0 at $\alpha=.01$. There is not an equal preference for all five sections.

15.75 (a) $\overline{X} = (0+3+4+...+0+2+3)/30 = 58/30 = 1.93$

(b) $P(X=0) = \frac{1.93^0 \, e^{-1.93}}{0!} = .145 \quad P(X=1) = \frac{1.93^1 \, e^{-1.93}}{1!} = .280$

$P(X=2) = \frac{1.93^2 \, e^{-1.93}}{2!} = .270 \quad P(X=3) = \frac{1.93^3 \, e^{-1.93}}{3!} = .174$

$P(X=4) = \frac{1.93^4 \, e^{-1.93}}{4!} = .084 \quad P(X=5) = \frac{1.93^5 \, e^{-1.93}}{5!} = .032$

(c) To find the expected totals for each group, we multiply each probability by 30, yielding 4.35, 8.40, 8.10, 5.22, 2.52, and 0.96. We must combine 0 and 1 together and 3 or more as a category. To test H_0: the distribution is Poisson, the actual counts are 6 0's, 5 1's, 8 2's, 8 3's, 2 4's, and 1 5's.

$\chi^2 = \frac{(11-12.75)^2}{12.75} + \frac{(8-8.10)^2}{8.10} + \frac{(11-9.15)^2}{9.15} = 0.62 < \chi^{2*} = 3.84$, based on

3 − 1 − 1=1 df. Do not reject H_0 at $\alpha=.05$. The distribution does not differ significantly from the Poisson distribution.

15.77 For H_0: The data fits a Poisson distribution, we first need to estimate λ, with $\overline{X} = [1(4) + 2(8) + ... + 9(77) + 10(12)]/721 = 6.06$. Therefore, we have probabilities of .0023, .0141, .0429, .0866, .1312, .1590, .1606, .1390, .1053, .0709, and .0881 for the probabilities of 0, 1, 2, 3, 4 5, 6, 7, 8, 9, and 10 or more. Since we do not have X=0, we combine this with X=1, yielding a probability of .0164, so that expected totals are obtained by multiplying by 721, yielding 11.82, 30.93, 62.44, 94.60, 114.64, 115.79, 100.22, 75.92, 51.12, and 63.52. Therefore, we have
$\chi^2 = \frac{(4-11.82)^2}{11.82} + \frac{(8-30.93)^2}{30.93} + ... + \frac{(12-63.52)^2}{63.52} = 139.077 > \chi^{2*} = 15.5$, based on 10 − 1 − 1=8 df. Reject H_0 at $\alpha=.05$. The data does not fit a Poisson distribution

15.79 We are comparing the fit of a sample of data into the categories of a suggested distribution

15.81 We can use either test when we have only 1 df for the χ^2 test. When this occurs, $\chi^2 = z^2$.

CHAPTER 16

NONPARAMETRIC STATISTICS

16.1 For H_1: The average number of accidents differs among contractors, we have $n_1=15$, $n_2=12$, $T_A = 159$, $T_B = 219 = T$. $\mu_T = 12(28)/2 = 168$, $\sigma_T = \sqrt{15(12)(28)/12} = 20.49$, $z = (219 - 168)/20.49 = 2.49 > z^* = 1.96$. Reject H_0 at $\alpha=.05$. Contractor B has a higher average number of accidents.

16.3 For H_1: The average nonbillable time is lower for professionals with master's degrees than for those with bachelor's degrees, $n_1=13$, $n_2=12$, $T_A=161.5$, $T_B=163.5 = T$. $\mu_T = 12(26)/2 = 156$, $\sigma_T = \sqrt{13(12)26/12} = 18.38$, $z = (163 - 156)/18.38 = 0.41 < z^* = 1.65$. Do not reject H_0 at $\alpha=.05$. There is insufficient evidence to show that the average nonbillable time is lower for professionals with master's degrees than for those with bachelor's degrees.

16.5 For H_1: the product is effective, $n_1=7$, $n_2=8$, $T_1=67$, $T_2=53$, $T=67$, Reject H_0 if $T \geq T_{Table,7,8} = 71$. Do not reject H_0 at $\alpha=.05$. There is insufficient evidence to conclude that the product is effective.

16.7 For comparing the medians, the differences are .5, 0, .5, 1, .8, .6, .7, .1, so after we eliminate the 0, we have $T^- = T = 0$, and $T^+ = 28$. For H_1: There is a difference in the median rates between home-builders and financial services, at $\alpha=.05$, $T^* = 2$, so since $T=0 < 2$, we reject H_0 at $\alpha=.05$, and conclude that the home-builders have higher median rates than the financial services, agreeing with our result from Chapter 8. For comparing the modes, the differences are .5, -1, 1, 2, 2, 0, .3, and 0, yielding $T^- = 3.5 = T$ and $T^+ = 17.4$. For H_1: There is a difference in the mode rates between home-builders and financial services, at $\alpha=.05$, $T^* = 1$, since we had two 0's and $n=6$. Since $T=3.5 > 1$, we do not reject H_0 at $\alpha=.05$, and conclude that there is no significant difference in mode rates between home-builders and financial services, agreeing with our result from Chapter 8.

16.9 For H_1: the preparations differ in effectiveness, the differences are 5, 5, -7, 3, 1, -25, ..., -6, 4, -18, so $T^+ = 62.5$, $T^- = 108.5$, $T = 62.5$, n=18, $\mu = 18(19)/4 = 85.5$, $\sigma = \sqrt{18(19)(37)/24} = 22.96$, $z = (62.5 - 85.5)/22.96 = -1.00 > z^* = -2.58$. Do not reject H_0 at $\alpha=.01$. There is insufficient evidence to conclude a difference in effectiveness between the preparations.

16.11 For H_1: The additive increases mileage, the differences are .7, 0, 1, .3, 3.1, .7, 1.6, $-.1$, -1.4, and 3. With n=9, $T^+ = 38$, $T^- = 7$, T=7, and we reject H_0 if $T \le 8$. Reject H_0 at $\alpha=.05$. The additive increases mileage.

16.13 For H_1: The program is effective in increasing sales, subtracting after $-$ before, we have 7 positive differences and 3 negative differences, so X=7, and $z = (6.5 - 5)/\sqrt{10(.5).5)} = .95 < z^* = 1.65$. Do not reject H_0 at $\alpha=.05$. There is insufficient evidence to conclude that the new program is effective.

16.15 For H_1: Capital expenditures increases, there are 8 positive differences and 4 negative differences, so X=8, and $z = (7.5 - 6)/\sqrt{12(.5)(.5)} = 0.87 < 1.65$. Do not reject H_0 at $\alpha=.05$. There is insufficient evidence that capital expenditures increased during the second half of the year.

16.17 (a) To test H_1: There is a difference between the two groups, there are 15 negative differences, 1 positive difference, and 1 0, which is eliminated. For n=16, $z = (14.5 - 8)/\sqrt{16(.5)(.5)} = 3.25 > 1.96$. Therefore, we reject H_0 at $\alpha=.05$ and conclude that the migrant mean is higher than the longterm mean.

(b) This would be valuable for any city which attracts a large number of migrant workers.

(c) No assumptions are necessary.

16.19 For H_1: 1987 was a losing year, $z = (7.5 - 5.5)/\sqrt{11(.5)(.5)} = 1.21 < 1.65$. Do not reject H_0 at $\alpha=.05$. No, it was not necessarily a losing year.

16.21 For H_1: The changeover was successful, we have 3 days with more than 15 errors, 15 days with fewer than 15 errors, and 2 days with exactly 15 errors, which we do not use for our calculations. with n=18, $z = (14.5 - 9)/\sqrt{18(.5)(.5)} = 2.59 > 1.65$. Reject H_0 at $\alpha=.05$. Yes, the changeover was successful in reducing the mean number of errors.

16.23 For H_0: $\mu_I = \mu_{II} = \mu_{III}$, after ranking the observations from low to high, we have $T_I = 72$, $T_{II} = 146$, $T_{III} = 106$, $n_I = n_{II} = 8$, $n_{III} = 9$, n=25.

$$H = \frac{12}{25(26)}\left[\frac{72^2}{8} + \frac{146^2}{8} + \frac{106^2}{9}\right] - 3(26) = 6.64 > \chi^{2*} = 5.99, \text{ based on 2 df.}$$

Reject H_0 at $\alpha=.05$. There is a difference among companies.

16.25 For H_0: $\mu_A = \mu_B = \mu_C = \mu_D = \mu_E$, after ranking the observations from low to high, we have $T_A = 84$, $T_B = 108$, $T_C = 95$, $T_D = 60.5$, $T_E = 117.5$, $n_A = 5$, $n_B = n_C = 7$, $n_D = 5$, $n_E = 6$, $n=30$.

$$H = \frac{12}{30(31)}[\frac{84^2}{5} + \frac{108^2}{7} + \ldots + \frac{117.5^2}{6}] - 3(31) = 2.48 < \chi^{2*} = 9.49, \text{ based on 4 df.}$$

Do not reject H_0 at $\alpha=.05$. There is not a significant difference among firms.

16.27 For H_0: $\mu_1 = \mu_2 = \mu_2 = \mu_4 = \mu_5 = \mu_6$, after ranking the observations from low to high, we have $T_1 = 79.5$, $T_2 = 141$, $T_3 = 62$, $T_4 = 189$, $T_5 = 44$, $n_1 = n_2 = n_3 = n_4 = n_5 = n_6 = 6$, $n=36$.

$$H = \frac{12}{36(37)}[\frac{79.5^2}{6} + \frac{141^2}{6} + \ldots + \frac{44^2}{6}] - 3(37) = 24.66 > \chi^{2*} = 11.1, \text{ based on 5 df.}$$

Reject H_0 at $\alpha=.05$. Yes, there is a difference among regions.

16.29 For H_0: no difference among the three hospitals, we rank the hospitals across the services, yielding 231, 231, 321, 321, 321, 312, and 321, for rank totals of 19, 15, and 8. With k=3 and b=7. Therefore,

$$F_r = \frac{12}{7(3)(4)}[19^2+15^2+8^2] - 3(7)(4) = 8.86 > \chi^{2*} = 5.99, \text{ based on 2 df. Reject}$$

H_0 at $\alpha=.05$. There is a difference among the hospitals.

16.31 We need at least 5 observations for each group of the tested factor. To test the states, we need at least two more years of data. To compare the years, we need at least two more states.

16.33 To test the colas, $\mu_I=\mu_{II}=\mu_{III}$, first we rank the colas across each person, from low to high. Thus, we have 231, 321, 123, 321, 2.5 2.5 1, ..., 321, and 222. Now, we total the ranks for each group, yielding $T_I = 22.5$, $T_{II} = 20$, and $T_{III} = 17.5$, b=10, k=3.

$$F_r = \frac{12}{10(3)(4)}[22.5^2 + 20^2 + 17.5^2] - 3(10)(4) = 1.25 < \chi^{2*} = 5.99, \text{ based on 2 df.}$$

Do not reject H_0 at $\alpha=.05$. No, at this time, there is not a significant difference among colas.

16.35 (a) To test the networks, $\mu_I=\mu_{II}=\mu_{III}$, first we rank the networks across each time slot, from low to high. Thus, we have 321, 213, ..., 213, and 321. Now, we total the ranks for each group, yielding $T_I = 23.5$, $T_{II} = 17.5$, and $T_{III} = 19$, b=10, k=3.

$$F_r = \frac{12}{10(3)(4)}[23.5^2 + 17.5^2 + 19^2] - 3(10)(4) = 1.95 < \chi^{2*} = 5.99, \text{ based on 2 df.}$$

Do not reject H_0 at $\alpha=.05$. No, at this time, there is not a significant difference among networks.

(b) Sometimes(hopefully most of the time). If the tests agree all of the time, there would be no need to have both of them.

16.37 When we take these ranks just within the twelve companies, we get 9,3,1,7,4,12,10,2,11,8,5, and 6, compared with 1,2,3,4,5,6,7,8,9,10,11,and 12. Therefore, we have $d = R_x - R_y = -8, -1, 2, -3, 1, -6, -3, 6, -2, 2, 6,$ and 6, so that $\Sigma d^2 = 240$. Therefore, $r_s = 1 - 6(240)/[12(143)] = .16$. To test $H_1: \rho > 0$,

$$t = \frac{.16\sqrt{10}}{\sqrt{1-(.16)^2}} = 0.51, \text{ which is less than any t value based on 10 df. No, there is not}$$

evidence of agreement.

16.39 For after tax sales, the ranks are 7, 8, 1, 2, 5, 3, 6, 4. For number of employees, the ranks are 8, 2, 4, 3, 5, 6, 1, 7, so the differences are 1, -6, 3, 1, 0, 3, -5, 3, and $\Sigma(R_x - R_y)^2 = 90$, so $r_s = 1 - 6(90)/[8(63)] = -0.07$. For H_1: After tax return is correlated to number of employees, at $\alpha=.05$, $r_s^* = -.738$. Since $-0.07 > -.738$, we do not reject H_0 at $\alpha=.05$ and conclude that after tax return is not significantly correlated to number of employees.

For number of meetings, the ranks are 7.5, 5, 5, 7.5, 5, 2.5, 2.5, and 1, so the differences are .5, -3, 4, 5.5, 0, $-.5$, -3.5, and -3, and $\Sigma(R_x - R_y)^2 = 77$, so $r_s = 1 - 6(77)/[8(63)] = 0.08$. For H_1: After tax return is correlated to number of meetings, at $\alpha=.05$, $r_s^* = .738$. Since $0.08 < .738$, we do not reject H_0 at $\alpha=.05$ and conclude that after tax return is not significantly correlated to number of meetings.

For 1984 sales, the ranks are 7.5, 5, 1, 2.5, 5, 5, 2.5, and 7.5, so the differences are .5, -3, 0, .5, 0, 2, -3.5, and 3.5, and $\Sigma(R_x - R_y)^2 = 38$, so $r_s = 1 - 6(38)/[8(63)] = 0.55$. For H_1: After tax return is correlated to 1984 sales, at $\alpha=.05$, $r_s^* = .738$. Since $0.55 < .738$, we do not reject H_0 at $\alpha=.05$ and conclude that after tax return is not significantly correlated to 1984 sales.

16.41 (a) $\binom{10}{2} = 45$

(b) For H_1: ICU rankings are correlated to pediatrics, the differences are 1, 0, 0, 1, 1, 0, -1, -2, so $\Sigma(R_x - R_y)^2 = 8$, and $r_s = 1 - 6(8)/[8(63)] = .90 > r_s^* = .738$. Therefore, we reject H_0 at $\alpha=.05$ and conclude that there is a positive correlation.

(c) We need to have $|r_s| > .738$ for correlations to be significant. It turns out all correlations are significant except ICU/OB(r=.619), and ER with Output(.643), Surg(.690), OR(.690), OB(.714), Ped(.714), and Psych(.714).

16.43 H_1: Weekly Salary and Employees Supervised are correlated. The ranks for Salaries are 1,2,3,4,5,6,7,8,9,10,11,12,13,14,15. The ranks for employees supervised are 2,4,3,6,7,1,5,12,9,10,8,14,11,13,15. Therefore, the differences are $-1, -2, 0, -2, -2, 5, 2, -4, 0, 0, 3, -2, 2, 1, 0$, so that $\Sigma(R_x - R_y)^2 = 76$, and $r_s = 1 - 6(76)/[15(224)] = .86$.

$$t = \frac{.86\sqrt{13}}{\sqrt{1-(.86)^2}} = 6.07 > t^*_{.05,13} = 1.77.$$ Reject H_0 at $\alpha=.05$. Numbers of employees supervised has a positive relation to salary.

16.45 Ranking the stocks yields 1.5,3,1.5,7.5,6,5,7.5, and 4. Ranking the returns gives 7,4,1,8,6,5,3, and 2. The differences are $-5.5, -1, .5, -.5, 0, 0, 4.5$, and 2, so that $\Sigma(R_x - R_y)^2 = 47$, and $r_s = 1 - 6(47)/[8(65)] = .440$. To test H_1: Numbers of stocks and returns are correlated, $r_s^* = .738$. Do not reject H_0 at $\alpha=.05$. There is not significant evidence of a relationship between the number of stocks and the portfolio return.

16.47 Non-parametric. We are not sure what ratings such as 8, 9, or 10 mean, but we can compare them in a ranking test. We use the Wilcoxon Signed Ranks Test.

16.49 For H_1: Mean ratings differ between the keyboards, we first pair off the differences, yielding 1, -4, 7, 3, 6, 15, -2, -2, 2, 3, 6, 4, -2, 2, and 9. After ranking these differences in absolute value, low to high, we have $T^+ = 98.5$ and $T^- = 21.5 = T$. $\mu = 15(16)/4 = 60$, $\sigma = \sqrt{15(16)(31)/24} = 17.61$, so $z = (21.5 - 60)/17.61 = -2.17 < z^* = -1.96$. Reject H_0 at $\alpha=.05$. Keyboard I has higher mean keyboard ratings.

16.51 For H_1: Sales are positively correlated with the size of the advertisement, the sales rankings are 8, 6, 10, 5, 4, 9, 1, 7, 2, and 3. The Ad rankings are 9, 7, 10, 5, 3.5, 105, 8, 6, 3.5, and 1.5. The differences are $-1, -1, 0, 0, .5, 7.5, -7, 1, -1.5$, and 1.5, so $\Sigma(R_x - R_y)^2 = 113$. $r_s = 1 - 6(113)/[10(99)] = .32 < r_s^* = .648$. Do not reject H_0 at $\alpha=.05$. There is not sufficient evidence that ice cream sales are positively correlated with the size of the advertisement.

16.53 For H_1: There is a difference in the mean stock ratings, first we pair the ratings, yielding 1, 7, 3, −7, −2, 11, 40, 0, 3, 60, −100, and −3. We eliminate the 0 observation and rank the others in absolute magnitude from low to high, generating 1, 3.5, 4, 6.5, 2, 8, 9, 4, 10, 11, and 4, so that $T^+ = 42.5$ and $T^- = 23.5 = T$. With $\mu = 11(12)/4 = 33$ and $\sigma = \sqrt{11(12)(23)/24} = 11.25$, so $z = (23.5 - 33)/11.25 = -0.84 > -1.96$. Do not reject H_0 at $\alpha=.05$. There is not a significant difference in the mean ratings.

16.55 Our rankings within days are 231, 123, 321, 132, 123, 231, 123, and 123, so that $T_A = 12$, $T_B = 19$, and $T_C = 17$. To test H_0: All three chains have the same mean daily sales, we have k=3, b=8, and

$$F_r = \frac{12}{8(3)(4)}[12^2 + 19^2 + 17^2] - 3(8)(4) = 3.25 < \chi^{2*} = 5.99, \text{ based on 2 df. Do not}$$

reject H_0 at $\alpha=.05$. There is not evidence of a difference among the three chains.

16.57 (a) They have the same hypotheses, but different assumptions. The H test uses ranks while the ANOVA F test uses the actual observations.

(b) Both involve rank sums of independent groups. Both are non-parametric, but the Wilcoxon Rank Sum test is for only two groups, while the H test is for two or more groups.

(c) They have the same hypotheses, but different assumptions. The Friedman test uses ranks while the ANOVA F test uses the actual observations.

(d) Both involve rank sums from two factors. Both are non-parametric, but the Wilcoxon Signed Rank Sum test can only test two groups at a time, while the Friedman test is for two or more groups.

16.59 For H_1: The median exceeds 125, n=30, X=38, and $z = (37.5 - 30)/\sqrt{60(.5)(.5)} = 1.94 > 1.65$. Reject H_0 at $\alpha=.05$. There has been an increase in reservations.

16.61 To test H_0: $\mu_1=\mu_2=\mu_3$, we rank the colas across each week, yielding 321, 231, 321, 321, 321, and 321. Thus, the rank totals are 14, 11, and 5. With k=3, b=5,

$$F_r = \frac{12}{5(3)(4)}[14^2 + 11^2 + 5^2] - 3(5)(4) = 8.40 > \chi^{2*} = 4.61, \text{ based on 2 df. Reject}$$

H_0 at $\alpha=.10$. There is a difference among the sodas.

16.63 For H_1: Those in banking give a higher priority, we note that a higher priority means a lower ranking, since 1 is the highest priority. When we order the observations, we have 1, 1, 1, 1, 1, 2, 2, 2, 2, 2, 3, 3, 3, 3, 4, 4, 5, 5, 8, so the tied rankings are 3 for the 1 terms, 8 for the 2 terms, 12.5 for the 3 terms, 15.5 for the 4's, 17.5 for the 5's, and 19 for the 8. Therefore, we have $n_A = 9$, $n_B = 10$, $T_A = 99 = T$, $T_B = 91$, and the table value, $T^* = 111$. Since $99 < 111$, do not reject H_0 at $\alpha=.05$. No, there is not sufficient evidence to conclude that those in banking give a higher priority.

16.65 For H_1: Organism level has dropped during the one-year period, the differences are -3, 0, 5, 8, 3, -10, 14, 29, 11, 14, 13, and 10. We eliminate the 0 and rank the others from low to high in absolute magnitude, generating 1.5, 3, 4, 1.5, 5.5, 9.5, 11, 7, 9.5, 8, and 5.5, so $T^- = 7 = T$, and $T^+ = 59$, and $\mu = 11(12)/4 = 33$, and $\sigma = \sqrt{11(12)(23)/24} = 11.25$, so $z = (7-33)/11.25 = -2.31 > -2.33$. Do not reject H_0 at $\alpha=.01$. There is insufficient evidence to conclude that the organism level has dropped.

16.67 Because the test statistics fail to follow familiar distributions such as normal, t, F, or χ^2.

16.69 For H_1: The two newspapers differ in their ability to attract advertisers, the paired differences are 106, 12, 99, 93, -7, 174, 18, -5, and -10. The rankings are 8, 4, 7, 6, 2, 9, 5, 1, and 3, so $T^- = 6$ and $T^+ = 39$, so $T = 6$. With n=9, we have $T^* = 3$, so since $T = 6 < 3$, we do not reject H_0 at $\alpha=.02$ and conclude that no significant difference exists at this time between the newspapers.

16.71 No, the group with the lower total in the sign test might well have some high rank differences. Therefore, it is well possible to reject H_0 for the sign test, but fail to reject H_0 for the Pairs test, and vice-versa.

16.73 For H_1: the mean number of free computers has increased, we have 3 observations at 7, 13 greater than 7, and 5 less than 5. Eliminating the 3 observations at 7, we have n=22, so $z = (12.5 - 11)/\sqrt{22(.5)(.5)} = 0.64 < 1.65$. Do not reject H_0 at $\alpha=.05$. There is insufficient evidence to conclude that the average number of available terminals has increased.

16.75 For the sign test, we just need to count how many observations are on each side of the tested median. Using the binomial distribution, we expect half on each side. For the signed-rank test, we need to deal with rank totals from groups which may not have equal sample sizes, so it is not intuitively obvious what to expect.

16.77 For H_0: Interest Rate and Percentage in Arrears are correlated, we have the interest rates rank as 2.5, 6, 1, 7, 2.5, 4, and 5, while the arrears rankings are 2, 3, 6, 7, 1, 4, and 5. The differences are .5, 3, -5, 0, 1.5, 0, and 0, so $\Sigma(R_x - R_y)^2 = 36.5$, so $r_s = 1 - 6(36.5)/[7(48)] = .348 < r_s^* = .786$. Do not reject H_0 at $\alpha=.05$. There is not evidence of significant correlation between interest rate and percentage in arrears at this time.

16.79 No

16.81 We need to test groups pairwise using (a) the Wilcoxon Rank Sum Test (b) the Wilcoxon Signed Rank Test

16.83 Recall that $\sum_{i=1}^{n} i = n(n+1)/2$, and $\sum_{i=1}^{n} i^2 = n(n+1)(2n+1)/6$. Therefore, we have

$SS_x = SS_y = \Sigma i^2 - (\Sigma i)^2/n = n(n+1)(2n+1)/6 - [n(n+1)/2]^2/n = n(n+1)[\frac{2n+1}{6} - \frac{n+1}{4}] = n(n+1)(n-1)/12.$

$SS_{xy} = \Sigma XY - [(\Sigma X)(\Sigma Y)/n] = \Sigma XY - n(n+1)^2/4$

We have $\Sigma(R_x - R_y)^2 = \Sigma R_x^2 + \Sigma R_y^2 - 2\Sigma R_x R_y$, so $\Sigma R_x R_y = n(n+1)(2n+1)/6 - \Sigma(R_x - R_y)^2/2$ and $SS_{xy} = n(n+1)(2n+1)/6 - \Sigma(R_x - R_y)^2/2 - n(n+1)^2/4 = n(n+1)(n-1)/12 - \Sigma(R_x - R_y)^2/2.$

$r = \dfrac{SS_{xy}}{\sqrt{SS_x SS_y}} = \dfrac{n(n+1)(n-1)/12 - \Sigma(R_x - R_y)^2/2}{n(n+1)(n-1)/12} = 1 - \dfrac{\Sigma(R_x - R_y)^2/2}{n(n+1)(n-1)/12} = 1 - \dfrac{6\Sigma(R_x - R_y)^2}{n(n+1)(n-1)} = 1 - \dfrac{6\Sigma(R_x - R_y)^2}{n(n^2 - 1)} = r_s$